'Consciousness is on. . . . . . . Fascinating'
Mail on Sunday

'The idea of the self as a relatively closed system is a delusion that has often conferred advantage, but is now a dangerous trap. Moving through difficult science with valuable clarity, Oliver tells us why . . . [a] timely, challenging book'
Richard Kerridge, *Guardian*

'Ambitious . . . it has several important messages, many of which need to be more widely understood . . . a thought-provoking and worthwhile read'
Tom Chivers, *The Times*

'A vision of a connected society – one that is conscious of its dependency on nature – is a glistening beacon in a gloomy ecological present. In an era that will be defined by its planetary action, Oliver's argument is timely and thorough'
Geographical

'Humans are less discrete entities than mash-ups of microbiota and shifting beliefs, declares ecologist Tom Oliver in this rich, intriguing book. We are, he shows, so interfused with the environment that all life might be seen as a web of genes, and all minds a web of memes'
Nature

'Interweaving the natural sciences with case studies from neuroscience and psychology, *The Self Delusion* assembles a compelling thesis . . . It's easy to see why he has won awards for communicating science to a general audience'
Sunday Business Post

'Heraclitus famously said that no man can step in the same river twice. Oliver has the science to back up the philosophy'
The Tablet

'*The Self Delusion* is a book of wonders. It articulately explores the infinite web of connection that humans have with one another, as well as with those that lived before them and those yet to be born. But Oliver also takes us beyond the body to tease out our many connections with the world around us and far, far beyond it. How can we really be individuals when trillions of atoms from the farthest reaches of the galaxy can be found in our bodies? A timely, fascinating and quite brilliant book'
Vybarr Cregan-Reid, author of *Primate Change*

'As a recovering individualist, I need to be reminded of the dangers of the self delusion and the benefits of dispelling it, and this book is a fascinating and compelling presentation of the scientific evidence'

Michael Foley, author of *The Age of Absurdity*

'As Tom Oliver takes us through this enlightening tour of our interconnectedness, from microscopic interactions to our collective cultural mind, he mercilessly dissects the very notion that our cherished individuality exists at all. Entertaining and thought-provoking, this book offers an urgently needed revaluation of our place in the world and what the next steps in our evolution might be'

Nathan H. Lents, author of *Human Errors*

'Until you've read Tom Oliver's delightful *The Self Delusion* you'll never have guessed that, from the ground up, you and your self-identity are constructions, built like an Arcimboldo painting, but of cells from many sources, neurons, ideas and finally connections to others. Read this book for a compelling way of thinking about how and why the 'you' that you see when you look inside yourself arises, and its place in the universe'

Mark Pagel, author of *Wired for Culture*

'Tom Oliver tells a compelling story, firmly rooted in biological evidence, that will make you think differently about yourself and your relationship with the world around you'

Sir Prof. Ian Boyd, Chief Scientific Advisor, UK Department for Food, Environment and Rural Affairs (2012-2019)

'If there was ever a book that connected us to current world it's this one. From how our bodies are made, the life of bacteria and its transmission between us, to how we create identity and what that means, and our connection and relationship to the natural world. Tom Oliver takes us on a revealing and important journey'

Alan Moore, author of *Do Design: Why Beauty is Key to Everything*

'*The Self Delusion* is a captivating and important book. Oliver presents an eye-opening scientific and philosophical discussion about the illusion of self and the tragic errors that stem from excessive self-centeredness'

Bill Sullivan, author of *Pleased to Meet Me: Genes, Germs, and the Curious Forces That Make Us Who We Are*

Tom Oliver is a professor at the University of Reading, leading their Ecology and Evolution research group. He is a prominent systems thinker, advising both the UK government and the European Environment Agency. He is a frequent contributor to broadcast media and regularly gives talks on environmental science to general audiences. He has published more than eighty scientific papers in world-leading interdisciplinary journals and won two first-place prizes for essays communicating science to a broader audience. His writing has appeared in the *Guardian*, *Independent* and *BBC Science Focus* and he regularly gives talks on environmental science to general audiences. *The Self Delusion* is his first book.

THE
SELF
DELUSION

The Surprising Science of
Our Connection to Each Other
and the Natural World

TOM OLIVER

First published in Great Britain in 2020 by Weidenfeld & Nicolson
This paperback edition published in 2021 by Weidenfeld & Nicolson
an imprint of The Orion Publishing Group Ltd
Carmelite House, 50 Victoria Embankment
London EC4Y 0DZ

An Hachette UK Company

1 3 5 7 9 10 8 6 4 2

Excerpt at the beginning of Chapter 11 from *Reasons to Stay Alive* © Matt Haig, 2015,
published by Canongate, Edinburgh. Used with permission.

Excerpt at the beginning of Chapter 11 from *Cartographies of Silence* from *The Dream of a
Common Language: Poems* © Adrienne Rich, 1974–1977, published by W. W. Norton &
Company, New York. Used with permission.

Excerpt at the beginning of Chapter 18 from *Angels and Demons* © Dan Brown, 2001,
published by Corgi of Penguin Random House, UK. Used with permission.

ISBN (Paperback) 978 1 4746 1176 3
ISBN (eBook) 978 1 4746 1177 0

Typeset by Input Data Services Ltd, Somerset

Printed and bound in Great Britain by Clays Ltd, Elcograf S.p.A.

www.weidenfeldandnicolson.co.uk
www.orionbooks.co.uk

Dedicated to the legions that are our antecedents

How connected are you?

Follow this link to find out:
https://tinyurl.com/u94zyf9

Contents

Introduction

As a child, I was eager to believe in the supernatural. It seemed intuitive to me that with enough practice I could master the ability to move objects with the power of my mind. I consumed books written by Indian yogis on how to levitate and invented secret phrases, believing these could help me save penalties in football if I imbued them with enough willpower. These inclinations continued, dare I say, until I was part way through my undergraduate degree. In a zoology practical, I investigated the turning behaviour of insects, observing them walk along thin lengths of wood, about the width of an ice cream stick. At the end was a T-junction where they were forced to turn left or right. The aim was to demonstrate how bugs that chose to turn left the first time, would next time turn right — a corrective behaviour allowing them to move in a broadly straight line overall. As I carefully watched them, I wondered whether I could harness the power of the human mind to influence the direction they initially turned. I even speculated that other species groups such as ants could be more easily influenced because, as social insects, they are sensitive to subtle cues from their many workmates. Maybe the electromagnetic fields of our human brains could interact with those of the ants to influence their behaviour and, if my brainwaves alone were

1

insufficient, what if the experiment were placed in a football stadium with thousands of people concentrating on influencing their turning. Might it just work?

Bizarre as the idea was, in my defence it was not 'anti-scientific' – I proposed to test it with a sound scientific experimental methodology. I had a clear hypothesis: that somehow the electrical brainwaves of humans could interact with those of ants and influence their turning decisions, and I was conceiving experiments to falsify or support my hypothesis. But while it was not anti-scientific, it was just probably not sensible given what we know about human brainwaves and ant behaviour in the twenty-first century. Many people would, from the start, discount the idea based on 'common sense' that mind-over-matter is nonsense, but I might have countered this by saying those people are not deeply questioning how the world works. I was refusing to unthinkingly accept the common wisdom and trying to challenge the status quo. Fortunately, I ended up picking a project that was less likely to be a complete waste of time.

Fast forward a few years, and I learned that my tendency to intuitively believe in mind over matter and, more generally, in the supernatural, was an inherent 'cognitive bias' – a systematic failure in rational judgement. Our human minds evolved this way; an ability to identify patterns of causality in a noise of input data is highly adaptive, even if it means we sometimes come to false conclusions. Even babies have this ability for self-deception, finding agency where none exists. Though false, these supernatural beliefs could still be beneficial if they helped early human groups to work together. Believing in a supernatural god who promised to reward cooperation and punish cheats helped people to work together successfully in closely knit groups.

From our tantruming toddler years to narcissistic adolescence, we eventually take stock one day to find we have reached the calmer shores of adulthood. Humans tend, mostly, to become more rational and less self-centred as they grow older. It is clear from recent neuroscience and psychology studies that our mind matures slowly and, as late as our third decade of life, the frontal lobes of the brain continue to develop higher reasoning capabilities. In a poetic way, the development of an individual human throughout their lifetime reflects in microcosm the long evolutionary journey of the human mind from early hominid to modern-day *Homo sapiens*. The primeval parts of our brain, the circuitry for the fear and aggression responses and the tendency to believe in supernatural agency, were dominant in early humans, and these are phases we pass through in our early life stages – from toddlers to adolescents. In later phases of our life, as fully developed adults, we have access to the higher cognitive reasoning that is the unique tool of modern *Homo sapiens*. Our species has evolved through biological adaptation and culture to develop a rationality which we have used to dispel myths, edging ever closer to seeing the objective truths of the universe. A significant milestone in the cultural evolution of human minds was the acceptance that the Earth is not the centre of the universe, the so-called Copernican revolution. However, we have one more big myth to dispel: that we exist as independent selves at the centre of a subjective universe. You may feel as if you are a discrete individual acting autonomously in the world; that you have an unchanging inner self that persists throughout your lifetime, acting as a central anchor-point with the world changing around you. This is the illusion I seek to tackle with this book, where I hope to convince you that we urgently need a Copernican-like revolution of human identity.

There are over seven billion humans on the planet as I write. Add to this all the humans who ever lived, or who will ever live, and ask does each one of those humans have their own universe that revolves around them – do they each exist as an independent centre-point? Or, are we all part of one con-nected, objective reality, influencing each other so strongly that the proposition we are independent entities is simply a defunct idea? To answer this question, we need to take a voyage into the centre of the human being. Do we even have a centre, and if not, what will we find? Like the characters of a Jules Verne novel who journey to the inner core of the Earth, we will also need a vessel to explore inside ourselves. That vessel is our imagination; fuelled with scientific facts, it will help us to explore our inner space. On our journey together we will travel through multiple dimensions: the physical body, our mental spheres and our social spheres. Then you must draw your own conclusion. Mine has been the realisation that supernatural powers do not exist, but also, surprisingly, that individual humans do not either: we are seamlessly connected to one another and the world around us. Our independence is simply an illusion that was once adaptive but now threatens our success as a species. A fundamental shift in self-perspective is the next step in our continued evolution as humans. There is a real urgency, because there are stark consequences of remaining stuck in our individualistic and atomistic mindsets, as we will discover.

It is time now to dismantle the illusion of individual human centredness, time to make the psychological transition we urgently need to solve the pressing environmental and social problems of the twenty-first century. It is time to conquer our 'self delusion'.

To begin, imagine that everything you have ever believed is a lie. Like a character in a movie, you have been living in a world of illusion. But unlike a Hollywood film there is no nefarious mastermind intent on keeping you trapped in a dreamlike state while selling off your body parts, and Keanu Reeves will not break into this dream and help you escape. It is up to you to examine your core beliefs about the world and to detach yourself from this illusion, to see reality with a fresh perspective.

Consider this statement: 'I am a man with brown hair and brown eyes who lives in the town of Wallingford.' Which part of this seemingly innocuous statement is untrue? Perhaps any of my gender, appearance or home address, you might say. But I assure you, you will find me looking as described if you visit and knock on my door (please don't, though). The falsehood in this statement lies at the start: an inoffensive, easily missed pronoun – 'I'. Behind this tiny letter lies a world of fantasy, the seed of an illusion. Consider, for a moment, who or what you mean when you refer to yourself as 'I'? Perhaps it feels as if this 'I' is the core of your character, the familiar seat from which you observe, interpret and initiate action in the world. Yet, it turns out we are programmed to think this. Like toy soldiers designed with limb mechanisms to move in a certain way, our brains are structured to create an illusion called 'I'. We have little choice over this programmed fantasy, although certain philosophies and cultures promote a stronger illusion of individuality, while others help us to perceive a deeper reality. In the modern globalised world, overcoming this self delusion is becoming increasingly essential for a whole host of important things like personal happiness, justice and maintaining a habitable global environment. No small thing, then.

But let's not get carried away with the idea of complete self-annihilation. The 'I' is also a survival tool. As essential as a penknife in the woods, it is part of our evolutionary toolbox that has allowed us to thrive as a species for over 200,000 years. Without some sense of individual identity, we would not be able to plan, motivate and direct our lives. However, just as certain behaviours, such as voraciously seeking out fatty foods, helped early humans to survive in the wild, our biological survival mechanisms do not always equip us in an optimal way for the modern world. We now realise that our fondness for fats leads to obesity in environments where calories are abundantly available, and we are gradually learning to use our rationality to overrule innate urges to binge on fatty food. In a similar way, the illusion of an independent 'I' has become maladaptive, leading to problems in an increasingly connected and globalised world. Despite overall improved quality of life, personal happiness is heading in the wrong direction. We are facing a mental health epidemic with a growing frequency of anxiety, depression and self-harm. Through selfish overconsumption we are destroying the natural world and using non-renewable resources at an accelerating rate. Pollution and the spread of antibiotic resistance are set to undermine the advances in human health of recent decades, while ethical dilemmas loom on the horizon as a result of climate change and mass human migration. I will explain how our sense of self-identity underpins these global issues, why they are becoming more severe and how we can get them back on track.

Reforming our individualistic self-perspective cultivates the mindset necessary to address these global problems. Their root cause is that we perceive the human condition as one of multiple independent entities vying for individual success, rather than all of us, being not only equal, but also deeply interlinked in connected systems. These personal views of the world become collectively enshrined in our institutions, and as long as these institutions that manage the environment, economy, justice and health continue to be based on our flawed logic, they will remain incapable of tackling the big problems that need solving for our species to survive. Fixing these problems starts with us. A first step in reforming our self-identity is to take a deep dive beneath the troublesome veil of illusion which gives us a sense of being an independent 'I'.

We tend to think of ourselves as discrete entities, as somehow separate and distinguishable from our surroundings, yet in many aspects – from our physical bodies to our brains and minds – we are deeply linked to the world around us. So connected, in fact, that an outsider objectively examining us – let's say an intelligent alien who is not susceptible to the same delusions of identity – might not be able to distinguish unique entities. Findings across a wide range of scientific disciplines increasingly support the idea that the central, discrete 'I' we obsessively nurture, protect and talk to throughout our lives is just an illusion.

Our body is a key part of our identity, yet most of the estimated 37 trillion cells that make up these bodies have but a short lifespan of days to weeks, so there is a near continual turnover of material. New molecules continually flow through us, derived from atoms from the furthest reaches of the universe, and which have also made up the bodies of countless other plants and animals before us. And since our bodies are essentially made anew every few weeks, the material in them alone is clearly insufficient to explain the persistent thread of an identity. Furthermore, most of the cells in our body are not even human: we contain more bacterial cells than human cells. Moreover some of these have the ability to influence our moods and manipulate our behaviours, further detracting from our supposed autonomy.

If not the materials in our body, what about the DNA instructions that code for its design, perhaps these comprise our unique identity? Just like the molecules that make up our bodies, our genetic code flows so fluently through – and between – the branches of the tree of life, that it is more like one great networked cloud computer program. Our bodies harbour a small subset of that code, cut and pasted into a transient entity.

If our DNA code does not comprise our unique identity, what about our minds: these are surely our own? Advances in psychology and neuroscience suggest that we have no unchanging, independent identity. Instead, we are a bundle of beliefs and self-reflections in constant flux. Our identity is contingent on the time of day, where we are and who we are with. Our perceptions are filtered by our consciousness, which is itself a product of those perceptions, and so our self-identity is a continually evolving product of the environment we are immersed in. This environment is hugely determined by other humans. Indeed, as humans we are grand architects of our environment. We have achieved this by being the most mutualistic species on earth. Try considering a simple man-made object that is close to you now. Its creation was contingent on the cooperative actions of hundreds, if not thousands, of humans across continents and over hundreds of years. Beyond the creation of these objects, our combined human endeavour contains the spoken and written cultures that so fluently cross the blood–brain barrier into our minds and unavoidably determine the way we think. To consider ourselves to be sovereign individuals is a deeply misplaced belief.

Unfortunately, we often struggle to comprehend our interconnectedness to others and the world around us. We struggle to see the bigger picture of our selfhood because we suffer from a form of blindness that we might loosely call an 'individualistic perspective', or more critically a 'self delusion' when we recognise its harmful aspects. The technical definition of a delusion is, according to the Diagnostic and Statistical Manual (DSM IV, the professional psychiatrist's bible of personality disorders), a false belief that is firmly sustained despite what almost everyone else believes and despite incontrovertible evidence to the contrary. We might ask why their definition makes a delusion contingent on what everyone else believes. Under the DSM IV definition, social norms usurp absolute truth and can make a blatantly false belief non-delusional, as long as everyone shares the same view. One can only speculate this caveat was added so psychiatrists didn't end up diagnosing the vast majority of the population with a mental disorder and getting themselves in a whole lot of hot water. However, it would seem more rational to label any belief as delusional if it is objectively untrue irrespective of social norms. As a thought experiment, imagine a future world where only a few people survive under the guidance of a cult leader who they all believe is Elvis Presley reincarnated. Would those people be deluded even though they all share this same belief together? If your answer is, like mine, that they would surely still be deluded, then you would probably also agree we should base our definition of delusion on the contravention of objective truths. We humans have suffered many such delusions over our cultural histories. We were once deluded about supernatural beings, we were deluded about a father-like God figure in the sky, and we are deluded that we exist as independent autonomous entities.

We are like a thread in a tapestry that is unaware of the majesty of the whole interconnected piece. This blindness is to an extent innate, but it has also been reinforced by our human culture. There exists a world beyond our senses, consisting of very small things, like the thriving diversity of bacteria and viruses with which we share our body, and of large-scale expansive processes, like the consequences of our actions that ripple across the entire globe. Just as a blind person can learn new ways to know the world they cannot see, so too can we learn to know the world through science. Technology and the scientific process are revealing a world beyond our limited human senses, but to fully 'see' this world we need to piece together the facts, like pieces in a jigsaw puzzle, to build a new perspective on the world and our place in it. Here, imagination is crucial. Just as novels help us to experience situations we have never encountered before by putting us into someone else's shoes, so too can imagination help us comprehend a world beyond our senses. Only with a vivid imagination can we synthesise scientific facts into a more accurate worldview that reveals our deep interconnectedness. Only with imagination can we see beyond an individualistic perspective and dispel the delusion of a discrete 'I'.

Coming back to the aliens observing the earth from space, would they really see a single *you* as a discrete physical body or, taking an objective science-based view (as intelligent spacefaring aliens might), would they rather see all the organisms on earth as part of a single body of life – a 'web of genes' – between which molecules flow. And would they observe *you* as a single consciousness, or would they look across all humans to see one great interconnected 'Mind' composed of linked sub-units between which ideas continually flow – 'a web of memes'? Encouragingly, this 'systems science' perspective adopted by the intelligent aliens is also increasingly common in our earth-based science. There is a slow tectonic shift towards holistic, systems-based thinking in physics, biology, social sciences, economics and many other disciplines. Using the systems approach, emergent properties can be observed that are hidden when simply analysing individual components.

Perhaps surprisingly, other areas of literature, including some religious and philosophical texts as well as poetry, have long recognised the deep interconnectedness of phenomena and have tried to describe experiences of unity, but the precision and evidence-based approach of science has been lacking. Although systems thinking is now becoming more mainstream in science, there has been little attempt so far to synthesise the implications of these approaches for humans and our place in the world. The outcome of any synthesis is greater than the sum of the parts. By reading this book I hope you will be able to draw together new and different perspectives to gain a bigger picture of the human condition and, as a collective, we will be in a better position to tackle some of the problems we face as a species, as well as becoming happier and more carefree.

Of course, comprehending a truth theoretically is different to integrating it into our deepest beliefs and attitudes, and so changing how we act in the world. To do that, we need to overcome deeply ingrained habits of thinking that we have been reinforcing most of our lives. Even when we mentally grasp the truth of our deep interconnectedness, we are often rapidly drawn back into our mental cockpit to see things with respect to a central and supposedly distinct 'I'. Have you experienced times when you have been anxious about something and catch yourself in the sudden realisation that you have been ruminating self-centredly for several minutes? How does this feel in contrast to moments when we interact with others, experiencing a deep sense of compassion, or when we are immersed in the natural world? These are wonderful, albeit transitory, moments of connectedness. Then we turn away from the other person, or head back indoors, and the feeling rapidly evaporates as trivial, self-centred concerns once again dominate our thoughts. The tendency for our minds to snap back to the perspective of isolated individualism is similar to what happens with an optical illusion. The Muller-Lyer illusion comprises parallel lines with a set of arrows at their ends. You probably know the one, where one line ends with inward-pointing arrows, the other line with outward-pointing arrows. Intuitively, the two lines seem to be of different lengths; the line capped with outward-pointing arrows appears to be shorter. Though we can grasp theoretically, and with some mental effort perceive, albeit briefly, the reality that the lines are the same length, when we lose concentration our perspective snaps back to its default view and we suffer the illusion once more. In a similar way, our minds are programmed through biological evolution – and reinforced by cultural evolution – to snap back to the illusion that we exist as autonomous discrete individuals. To truly dispel this illusion, we need to do more than just acknowledge it theoretically; we need to proactively overturn our deeply held mental models of the world. If we are successful, then it may catalyse a seismic shift in society. And it would not be too soon: we are at a critical crossroads as a species, where we must rapidly reform our mindsets and behaviour to act in less selfish ways. Otherwise, we will face increasingly hostile conditions that threaten not only our personal health and those of our families, but even the very future of our species.

We heard earlier how the development of the human mind towards greater rationality throughout our lives is like a microcosm of the long evolutionary journey of the human brain. In a similar vein, the transformation of our personal mindsets towards a more expansive, networked sense of self-identity may ultimately be reflected, in macrocosm, in a broader seismic cultural shift in society. Simultaneous changes happening in multiple people can lead to a domino effect preceding a rapid evolution of culture. So, although the negative trends in health and global sustainability may seem gloomy, we shouldn't forget that societal change can occur very quickly. A transformation to a more networked human identity may be just around the corner.

These ideas seem a long way from the supernatural beliefs of that young boy who tried to steer ants with his mind. Science tells us that particular feat cannot be achieved, not least because there is not even any independent mind to achieve it in the first place. As a famous economist once said in response to accusations over how he was changing his mind over an important matter: 'When the facts change, sir, I change my mind! What do you do?!'[1] In this book, I will provide facts that reveal we are not sovereign individuals, but part of a deep interconnected universal network. I ask you to take the quote above more literally than it was probably first intended: consider *changing* your mind in light of the facts – change your mental perspective, loosen your grip on the illusion of an independent 'I' and open your eyes to the hidden connections all around you. I believe you will find it opens the door to a more exciting, happier and fairer world.

PART ONE

OUR INTERCONNECTED BODIES

To see the World in a Grain of Sand,
And a Heaven in a Wild Flower;
Hold Infinity in the Palm of Your Hand;
And Eternity in an Hour

William Blake, 'Auguries of Innocence'

All things by immortal power,
Near or far,
Hiddenly,
To each other linked are,
That thou canst not stir a flower
Without troubling of a star

Francis Thompson, 'The Mistress of Vision'

We shall walk together in this path of life, for all things are part of the universe, and are connected with each other to form a whole unity   Maria Montessori, *To Educate the Human Potential*

1

You literally soaked up the atmosphere

Imagine the scene: it's a stormy night in Chicago with the howling wind and rain dashing against the Victorian windowpanes of the Old Infirmary. A flash of lightning reveals the form of a man hunched over a table with a bloody scalpel in hand. He shuffles across the room to the boom of thunderclaps, takes up a pen and starts to write: 'The subject was a white male 46 years of age, 53.8 kilos in weight, and 168 cm tall. Death was due to skull fracture as a result of a fall.' He stops writing and turns to glance over his shoulder at a table supporting a shape covered in a sheet, before continuing with a feverish excitement in his spidery scrawl: 'Dissection of the cadaver to obtain the various organs and tissues desired for study . . . performed on a stainless steel table covered with a polyvinyl sheet.' Soft tissue samples prepared for analysis by first being diced in a porcelain bowl with stainless steel knives.'

This macabre scene sounds like it might be from some kind of gothic horror fiction. You may be surprised, and a little worried, to know that the quoted text is taken word for word from a scientific journal paper in 1953 by Dr R. M. Forbes from the University of Illinois.[1] What kind of strange experimental dissection was this? The aim of this seemingly gruesome study was, in fact, to determine the composition of the adult human body through chemical analysis.

The question 'what are we made of' has troubled us for millennia. In the ancient Far East, Hindu and Buddhist scholars believed the substance of humans and everything else on earth was a mix of five elements: fire, earth, air, water and 'akasha' – the first four accounting for the material world and the fifth element accounting for the void beyond it. Across the globe, the Ancient Greeks settled on a similar idea, with the fifth element named 'aether'. Aristotle suggested poetically (but on rather flimsy evidence) that the stars must be made of this unchangeable and heavenly substance. These five elements were presumed to combine to make the human body, which Hippocrates described as being composed of four 'humours': yellow bile, black bile, blood and phlegm. Fantastic as it might seem, this theory held some sway for hundreds of years right up until the advent of modern European medicine in the nineteenth century. Perhaps its tenacity rested on the observation that blood in a glass container separates into four layers: a dark clot of black bile at the bottom, then red blood cells, then white blood cells they called phlegm, and finally a yellow serum at the top, which they called yellow bile. An imbalance of these humours was thought to lead to ill health, with an excess of black bile revealing susceptibility to deep depression. Some of the words in the English language still used today reflect this medical attribution. Melancholy is derived from the Greek *melankholia* meaning black bile, and phlegmatic describes someone who is calm and stoical.

Although the balance of humours in your body was thought to underpin your personality, there was also scope for changing this balance, for example, through the foods you ate. In Shakespeare's *Taming of the Shrew* a livid Petruchio is served overcooked meat and shouts to his wife, 'I tell thee, Kate, 'twas burnt and dried away and I expressly am forbid to touch it for it engenders choler, planteth anger and better 'twere that both of us did fast!' If a person became ill, terrible remedies were prescribed, such as laxatives or the letting of blood to restore the imbalance of the bodily humours. Such practices continued for hundreds of years until around the nineteenth century, when a more rigorous approach to science meant such ideas were eventually discredited.

So what did modern science find to replace the theory of the 'four humours' and explain the composition of the human body? Returning to our friend Dr Forbes, his macabre study may be able to shed light on the question. The results from his experiment are remarkably close to many later studies using more refined techniques. The composition of the human body is roughly as follows: oxygen (65 per cent of the total mass), carbon (18 per cent), hydrogen (10 per cent), nitrogen (3 per cent), calcium (1.5 per cent) and phosphate (1 per cent). The remaining 1.5 per cent is made up of a large number of other elements, including things like mercury, titanium and arsenic, in trace amounts. We are a cocktail of elements, and no yellow bile in sight. There was one thing the ancient medics got right, though: water (H_2O) is a primary component of our bodies.

Compared with all the fanciful theories about the constit-uents of our bodies, the reality seems perhaps rather mundane – we are comprised of the same commonplace elements that make up everything else in our universe. Dig deeper though and we may find that the commonplace is more extraordinary than we first think. Where did the chemical elements come from that are in your body right now? Where were each of the oxygen and carbon molecules before they found their way into making up your person? Well to start with, let's consider how many molecules we are actually talking about.

If we take oxygen, the most common element in our bodies, the average person in the world, weighing 62 kg,[2] contains just over 40 kg of oxygen. (It's perhaps surprising then that we don't all float around rather than walk, but the close attraction between water molecules make it denser than air.) To work out how many molecules there are in this substantial mass of oxygen, we need to go back to some school chemistry: to convert the mass of an element to the number of molecules it contains you first need to know its standard atomic mass. Dig out a periodic table and you'll see the symbol O, for oxygen, is accompanied by the value 15.999. This is the mass of one 'mole' of a substance (a mole is a constant quantity, a bit like a 'dozen', but bigger; a lot bigger in fact – it is roughly 6.02×10^{23} molecules and known as Avogadro's number). Dividing our mass of oxygen in grams by its standard atomic mass (15.999) and multiplying it by Avogadro's number gives an estimated 1.52×10^{27} oxygen molecules in the average human body. That is an exceptionally large number. Probably too big for us to imagine easily.

Here's a way to visualise it. Try to imagine the volume of the atmosphere around our planet. The earth's atmosphere actually has no definite upper boundary as it merges into outer space, but an accepted arbitrary definition is the 'Karman line' (named after a Hungarian–American engineer and physicist) which is 100km above sea level. The volume of the atmosphere between the earth and the Karman line comes to nearly 52 billion cubic kilometres.[3] Another very big number. Returning to the number of oxygen molecules in your body, let's imagine them exploding outwards and dispersing into the atmosphere so that there is an equal distance between all of them, producing a cloud of molecules surrounding the entire earth. How far apart would each oxygen molecule be?

They would be just 0.33 mm apart. Or, to put it another way, in every cubic metre of the atmosphere around the entire Earth there would be roughly 29 million of your oxygen molecules.[4] Imagine this dense fog of molecules stretching around the entire planet and then, in the blink of a geological eye, they are pulled together into one body – your body – right now. You have incorporated them into you with every breath you have taken throughout your life, with the food you have eaten and with the water you have consumed. The molecules have come from farms, rivers and springs from across the entire world, but what about before that?

Each molecule in our body has its own amazing history. One particular oxygen molecule may have circulated in the atmosphere for hundreds of years, swirling in eddies of wind that scoured the great deserts and the arctic tundra then, in a tropical storm, it may have plunged into the sea and continued its journey in a slow waltz among the currents of the great oceans. Another molecule may have passed between countless bodies – from fish to birds, plants and insects – before entering yours. It is not an exaggeration to say the molecules in your body have been recycled through the dinosaurs that walked the earth millions of years ago. All the organisms that have ever existed on the earth since the origin of life, approximately 3 to 4 billion years ago, are made of these same molecules, recycled in a never-ending cosmic dance. And each molecule has followed a different global path through time before ending up together in your body at this moment.

And before that? All the elements that make up our bodies were originally converted from hydrogen and helium, the two elements produced as a result of the 'Big Bang' at the origin of the universe. As stars formed, intense nuclear fusion processes deep inside their cores led to hydrogen and helium forging into a diverse range of elements. The atoms in our bodies are derived from these. Just as the American astronomer Carl Sagan suggested, we are literally made of star-stuff. However, we are not simply made up of atoms from stars in our own Galaxy, the Milky Way. A study by US and Canadian astronomers in 2017 found that over half of the atoms in our bodies have travelled from far off parts of the universe. Daniel Anglés-Alcázar and colleagues used computer simulations ('cosmological hydrodynamic simulations') to confirm that atoms spewed out by dying stars, as they become supernovae,[5] are picked up on streams of charged particles known as 'intergalactic winds', which travel between galaxies. Large galaxies like our Milky Way amassed half their matter from neighbouring star clusters up to a million light years away. As the authors describe succinctly, and somewhat poetically: 'intergalactic accretion feeds galaxies from the cosmic web'. So it turns out that atoms from the furthest reaches of the universe are gathered in our bodies. Quite a thought, that may provide an antidote for when we are feeling isolated: take a deep breath, inside your body is a meeting place for molecules from across the entire universe. Aristotle's claim that the stars were made of aether, which was also part of our bodies, turned out to be tantalisingly close to the truth.

Right now, trillions of these molecules are packed closely within your body, but this is just an infinitesimal blip in the life of the universe and, when your life expires, they will be released to continue their own journeys across the globe. As a dead body dries and decays, the molecules will leap forth again and will ultimately form the bodies of countless creatures to come. As the physicist Fritjof Capra suggests, our bodies do not really die but 'they live on and on again'.[6]

We started this chapter discussing the somewhat fanciful and fantastical theories of what makes up our bodies. In the face of aethers and humours, the true explanation that we are made from a set of chemical elements which also make up everything else in the universe may, at first, sound somewhat inane and dull. But a momentary meditation on this subject reveals the truth is far more awesome and exciting than we might first think. As the character Sherlock Holmes in Conan Doyle's novel says to his friend Watson, as they sit beside his fire in Baker Street: 'My dear fellow, life is infinitely stranger than anything which the mind of man could invent. We would not dare to conceive the things which are mere commonplaces of existence.' So it is with the origin of our human bodies; they are made up of molecules from across the universe and that once formed the bodies of countless others. Though it is quite beyond the scope of our day-to-day comprehension, we are fantastically connected to and intimately part of the Earth around us.

2

You are a work in progress

It is early morning on 4 April 1994 and particularly cold in the Sonoran Desert in Arizona. The silver moonlight glints like ice on the raised arms of huge saguaro cacti. A lone figure picks her way deftly around them, her shoes crunching on the cold sand. Up ahead lies a huge glass structure shaped like an Egyptian pyramid, lit up brightly from the inside. It is connected to a complex of other buildings like a great city of light surrounded by the dark desert.

The pyramid towards which Abigail Alling heads contains no less than 150 rainforest plant species. Unlike the air of the desert, which on this cold night is not far above freezing point, the night-time temperature is a balmy 22°C inside the super-insulated glass walls of the pyramids. Condensation streams down the walls, dripping onto the trees below. This rainforest biome is connected to others – a fog desert (where mist supplies the water for plants to survive), an ocean area and, in a large minaret-like structure overlooking the centre of the complex, the human living quarters.

Abigail approaches the pyramid and stands in front of it,

silhouetted against the triangle of light. Inside is the culmination of more than a decade of her life's work. She was involved in the design of the biomes, painstakingly selecting and planting the species within. In 1991, she was part of a team of six 'biospherians' on a pioneering scientific mission who would be sealed inside the glass complex for two years. Their mission: to become self-sufficient in the self-contained miniature world. The team would grow their own food and carefully monitor the atmosphere, minerals and water. The project was called Biosphere 2, in reference to Biosphere 1 being Planet Earth itself.

Abigail clenches her fists tight and starts towards the pyramid, breaking into a run heading for the two doors at the bottom of the structure. They are labelled with a warning 'Do Not Open Unless in Emergency', because doing this would expose the sealed unit to the outside air, compromising months of scientific research. Abigail grabs the handles and pulls. The doors crash open and a burst of warm, moist air smelling of soil and foliage rushes out towards her. Minutes later police cars swing into view and Abigail is arrested by federal marshals. She is later charged with criminal trespass, criminal damage and burglary.[1]

The strange behaviour of Abigail Alling might be viewed as particularly odd, given the years of dedicated work she devoted to the project. But, if you were to listen to her side of the story, she would explain how she was acting in the interests of her fellow scientific colleagues locked inside. The Biosphere 2 project was plagued with failures from the start, losing millions of dollars and forcing a new management team to be flown in.[2]

Locks on office doors were changed, the police called in and the previous senior staff were barred from accessing the site. Yet, the new team in charge had no proper knowledge of how to run the complex environment, and Abigail was concerned for their safety. As she explained to a colleague: 'I judged it my ethical duty to give the team of seven biospherians the choice to continue with the drastically changed human experiment ... or to leave ... It was not clear what they had been told of the new situation.'[3]

It turned out that maintaining a closed ecological system was hugely difficult. The atmosphere inside the Biosphere 2 biomes fluctuated wildly. High carbon dioxide levels meant that the health of the research team was compromised, and oxygen had to be pumped in. Many plant and animal species died, and some of those that survived became dominant and grew out of control.[4] Far from being a sealed unit, the doors had to be opened on a regular basis to bring in extra food. In the first mission, the biospherians were only 80 per cent self-sufficient in meeting their nutritional needs.[5] The team suffered from vitamin D and vitamin B_{12} deficiencies, meaning that extra calories and vitamin supplies had to be brought in. Towards the end of the first two-year mission, it was clear that artificial lights were needed as well as the introduction of predatory mites and biocontrol agents to control crop pests. Over the course of the project, the advisory panel of scientists, including NASA experts, quit over differences with the directors. At great expense, the ultimate goal to produce a closed ecological system had failed.

If the scientists had looked more closely into the workings

of Biosphere 1, the earth itself, they may have thought twice about embarking on their experiment. In the 1970s (more than two decades before), a scientist called James Lovelock proposed a hypothesis known as 'Gaia theory', which described the earth as a self-regulating, complex system helping to maintain the conditions for life on the planet. In Lovelock's theory, processes needed to maintain the earth's ecosystem (mineral cycles, ocean currents, gas exchange with the atmosphere) operate over vast global scales and strongly interact with one another. For the sceptical scientist, the name of the theory has slightly unfortunate connotations of sentience (indeed the original word *Gaia* is the name of the ancient Greek deity that personifies Mother Earth). It suggests a 'co-evolutionary' process between individual organisms and the environment, which maintains the stable conditions for life. According to Lovelock, when conditions on the planet, such as oxygen levels in the atmosphere, ocean salinity or global surface temperatures, deviate from the optimum for life, feedback mechanisms kick in to rebalance the system. The popularised version of the theory tends to paint the earth as some kind of 'superorganism', although the suggestion of a directed, purposeful agency behind organisational processes is an emphasis which its creator Lovelock denies in later writings.

Regardless of this dispute, Gaia theory has certainly highlighted one set of truths very well: the strong interconnectedness between earth systems such as the biosphere – all life on earth – the hydrosphere – all water on earth – and the lithosphere and pedosphere – all rocks and soils – which later research has repeatedly confirmed. A new discipline of earth systems science has developed around these planetary-scale interactions. In many cases, research results are surprising and challenge our governance systems, because the processes investigated surpass national boundaries. For example, cutting down forests in one area and replacing them with pasture or crops can lead to less rainfall in neighbouring regions. This means large forests like those in the Amazon and Congo basin which are dependent on high rainfall could be under threat, not only because of local deforestation, but also because of land management occurring in several South American or African countries.[6] The study of the formation of weather patterns over very large spatial scales gave rise to the popular term the 'butterfly effect'. This idea was initiated by the American mathematician Edward Lorenz, who found that tiny changes in the initial conditions of his computer simulations led to very different outcomes in terms of global weather patterns. Like many scientific discoveries he stumbled upon this by accident; he could not work out why he was getting very different weather outputs from repeated simulations, until he realised he had entered a certain input value as 0.506 instead of 0.506127. The tiny differences in initial conditions multiplied up and led to completely different predicted global weather patterns. Lorenz titled a subsequent scientific presentation 'Does the flap of a butterfly's wings in Brazil set off a tornado in Texas?', suggesting that something as small as a butterfly's flights in one location could affect the weather in a completely different country. Whether that is exactly true or not, it powerfully highlighted the interconnectedness of the whole planetary weather system.

Given the Earth's interlinked processes, it was perhaps naive to think one could simply seal off a small subset of the system, as attempted in Biosphere 2. No matter the detailed level of planning devoted to the regulation of the internal environment, the expensive glass biomes in the Arizona desert were doomed from the start (although one could argue that useful lessons were learnt during the process). Yet, where human technology is not yet fit to the task, what about nature, where millions of years of evolution has had the chance to innovate? Consider the bodies of mammalian species – these have evolved over millennia and represent excellent examples of self-regulating systems.

Our bodies are truly fantastic machines. While you read this, hundreds of complex feedback systems are operating to ensure your internal conditions, such as your core body temperature, blood pressure and pH levels, remain within optimal bounds for your continued survival. Despite the temperatures outside of our bodies varying wildly by tens of degrees as we walk in and out of doors, the internal body temperature rarely fluctuates by more than 0.5°C from the average (usually around 37°C in most people). Indeed, a change in internal temperature by more than 2°C causes amnesia and stupor when we become too cold (hypothermia), or nausea, vomiting, headaches and seizures when we become too hot (hyperthermia). Both can be potentially life threatening. The stakes are clearly high, so a whole range of highly effective feedback mechanisms kick in when our core body temperature starts to change. A region in the brain called the hypothalamus monitors blood temperatures, and if they rise too high, it fires electrical signals along nerve cells throughout your body, causing tiny blood vessels near the skin to dilate releasing heat from the skin's surface (and giving us a red and flushed appearance). Sweat glands are also activated leading to the excretion of water which, as it evaporates, cools you down. Conversely, if your core body temperature is at risk of becoming too low, blood is diverted away from the body's external edges and muscles are triggered into an involuntary repeated twitching, which generates heat (making us look pale and shiver awkwardly).

A whole range of other internal conditions are under similar feedback cycles. The kidney is a particular powerhouse of regulation, maintaining levels of chemicals such as glucose, sodium and potassium by selectively absorbing them from the bloodstream. Just as the role of the soils in Biosphere 2 was to remove toxic chemicals by pumping air through them, the kidney does a similar (and far more effective) job maintaining a healthy chemical balance in our bodies. As with temperature regulation, receptors in the brain control this balancing process, this time through the secretion of hormones which act as chemical messengers, rather than electrical signals. So if your blood pressure is dropping dangerously, the pituitary gland in your brain releases a hormone telling the kidney to take up sodium, secrete potassium, and increase water retention, thereby increasing your blood pressure to optimal levels. And it all happens without you realising it.

Any biology textbook can document many more such homeostatic mechanisms – evolved feedback processes allowing our bodies to regulate their internal environment. Our human cultures have then developed further upon these biological innovations. We have invented houses with home-ostatic heating and air conditioning systems to improve our survival and comfort. All this clever homeostasis should not be confused with the idea that we are *independent* of the external environment. A simple thought experiment will suffice – try holding your breath and see how long you survive!

Although they clearly have strong regulated internal condi-tions, our bodies are to some degree *open* systems – we depend on the regular input of energy and matter to survive. We need to replace the water we lose as we exhale and perspire, and we need sugars, carbohydrates and fats to build and fuel us, so we eat about thirty-five tonnes of food and drink about 31,000 litres of water throughout our lifetime.[7] To enable the release of energy from food, we need oxygen, obtained through the air we breathe – around 300 million litres of it over the course of our lives.[8] The oxygen in this inhaled air converts the energy stored in our food into a high energy molecule called ATP (adenosine tri-phosphate) that is the molecular currency used to continuously repair cells and power our muscles. At any one moment there is only about 250g of ATP in our bodies – enough to power only a single AA battery – yet, we create and use the chemical so rapidly that we convert our entire body weight of ATP each day.[9] These inputs of food, oxygen and water are very efficiently transferred into useful energy, but we still lose energy as heat and in the urine and faeces we excrete, which are the outputs to the system – energy and food in, waste and heat out.

Because our bodies are open systems linked to the envir-onment by flows of energy and matter continually passing through them, the barriers of this system (our skin, gut and lungs) need to be semi-permeable. They have evolved the ability to retain internal conditions (a steady temperature, pH, water pressure) quite different to our immediate environment, yet also allow the ingestion of the materials we need to fuel our bodies. Inside this semi-permeable body wall, a whole range of specialised organs (liver, heart, kidneys) work together to process the input materials, extract energy from them and deliver it to where it is needed. These organs regulating the constant flow of materials and energy are also very dynamic themselves – the cells they are made of do not last for their whole lifetime, so they need to be continually replaced.

In one episode of the well-known British TV comedy *Only Fools and Horses*, the character Trigger holds up a broom and proudly tells his friends he has had this broom for twenty years, then adds that in this time it has had seventeen new heads and fourteen new handles. His friends look confused; 'How can it be the same broom then?' one of them asks, to which Trigger responds by showing an old photograph of him holding the broom, as if this resolves the issue because the two brooms do look unarguably similar. Trigger may be the comedy fool of the TV series but perhaps he is right on this occasion. Is the broom defined by the molecules that made it originally, or is it something more than this?

A similar paradox has been raised many times over the centuries. The Greek essayist Plutarch tells the story of Theseus, the King of Athens, who had a ship in which every plank had been at some point replaced. In terms of its comparison with the original ship, Plutarch suggests philosophers were split in their opinions, with half maintaining the ship remained the same and the other half contending it did not. Our bodies present an almost identical paradox. The materials we ingest are continually used to rebuild our organs, because our cells die frequently. We might think of our body as a close friend or partner that is with us throughout our life, yet the average cell lasts only seven to ten years.[10] A team of Swedish researchers from the Karolinska Institute in Stockholm used carbon-14 dating techniques, similar to those used in palaeontology and archaeology, to age the different cells in the human body. They found that many cells are particularly short-lived. For example, gut lining cells live only five days, skin epidermis cells two weeks and red blood cells four months, while others last significantly longer – skeletal and other intestinal cells live around fifteen years. Cells that die before their owner are efficiently recycled by 'housekeeper' cells, passed in our faeces, or just shed from our skin into the air (humans are estimated to shed about 1 million microscopic particles every hour). So our bodies are very much like the Ship of Theseus, with nearly all of the original material regularly replaced.

In deep philosophical moments if we ask ourselves 'Who really am I?', it may be tempting to look at your body in the mirror and say, 'Well here I am, this is me!' Yet, like the Earth's ecosystems, our boundaries are porous with energy and matter flowing continually through them, while the component cells that make up our bodies are constantly recycled. If we are open systems interlinked with the outside world, can we really identify our body as being the source of an unchanging, independent 'me'?

3

Are you a human or a chimera?

Extracts from the diary of a mite:

Day 1: Feeling a bit headachy today. Our host, Jack, was at a nightclub dancing the night away, and we were out on the lash all night. Just fancy curling up in my hair follicle and munching some oily junk food today . . .

Day 2: Did a lot of walking last night. Feeling a lot healthier and happier now. It can get you down a bit, this place. The rest of them on this eyelid – my first, second and third cousins – are all right I guess. Don't get me wrong, they're a good bunch of mites and we have a laugh. But I don't want to spend the rest of my days with them. At some point in your life you want to settle down with someone special. There was someone once. I only saw her from a distance. Jack was at the cinema buying a ticket from the girl at the kiosk – a brunette with nice brown eyes and a great set of eyebrows. I was out on Jack's left eyelid, just hanging around and checking out some of the new film releases. Then I saw her, across the way. The prettiest mite I'd ever laid eyes on. The loveliest face and chelicera like petals. As Jack paid for his ticket, his eyes met the ticket girl's and they smiled at each other. At the same moment, I kid you not, at exactly the same moment, Däna (her name, as I later learned) looked from the ticket seller's eyelash across to me. Our eyes met. It was beautiful, and I felt for the first time in my life like I had finally come home. Then it was over. Jack had bought the ticket, we watched the film (a rubbish Vin Diesel movie) and made our way back on the cold night bus.

Day 3: Busy day catching up on some feeding. Not much to report. Keep thinking about Däna. Can't get her out of my mind.

Day 6: Jack is going on a date! He found a number on the back of his cinema ticket stub and gave it a call. They're meeting tonight. I'm so excited!

Day 7: Wow! What can I say? What a night. A perfect night. Jack went out for a meal with the brunette ticket seller, they were getting intimate and he leaned across for a kiss. Their faces touched and, as they did, I scuttled across from one eyelash to the other into Däna's open arms, all eight of them. We hit it off at once and since then we've been getting on a storm.

Day 15: Sorry I haven't written for a while. Däna and I are an item now. Baby mites!

The above might have happened, although perhaps (just

perhaps) the emotional life of eyelash mites is not quite as rich as portrayed here, but the events are not far wrong. *Demodex* mites live on the faces of nearly all of us and are able to transfer to the faces of others through close contact. They are around 0.2–0.4 mm in length and feed on the oils in our hair follicles, probably doing most of us no harm at all.[1] These tiny animals are just one part of the micro menagerie associated with our bodies. We are hosts to thousands of species of bacteria, fungi, protozoa and viruses, all in such dynamic abundance that we can truly be described as walking ecosystems.

The total number of bacteria that inhabit our bodies – our microbiome – is estimated to be around 38 trillion cells, slightly outnumbering the number of human cells.[2] In terms of the DNA these non-human cells contain, the ratio is more striking still. Humans have around 24,000 genes, whereas the total number of genes in the microorganisms associated with your body is thought to be around 2 million. Every surface of our bodies – from our faces to our armpits, from the insides of our mouths to the deepest depths of our guts – are covered in these microbes. For example, the Human Microbiome Project, a huge collaborative effort to quantify the number of microorganisms associated with us, found that we have over 1,000 bacterial species in our mouths, 440 in our elbow joints and 125 species behind our ears. As the scientific methods for observing this secret microscopic world steadily improve, we discover more – the more we look, the more we find – even in regions of our bodies we thought to be sterile; inside our lungs, our eyes, our brains, we are finding non-human life.

Some of these creatures are temporary companions – like

travellers sharing the road with us for a while, before departing. They may survive perfectly well outside our bodies in the wider environment, like the bacterium *Escherichia coli*, which occurs on foods like vegetables, as well as colonising our intestines and occasionally making us ill. Our microbiome is dynamic and changes throughout our lives, sometimes rapidly over the course of several days. If we change our diet, then the composition of our gut bacteria changes too, while the type of bacterial species on our skin is altered by soaps, skincare products and deodorants. Our general environment – how much time we spend outdoors, and who we live with – also affects the population of our microbiome. Researchers have found we share more bacteria in common with our housemates and our pets than strangers with whom we have little contact.[3]

To us humans, the changing state of our microbiome may seem quite rapid, in some cases with a full bacterial generation completed in twenty minutes, but to them, whose life cycles are much faster, the changing dominance between bacterial colonists are like long wars waged over several generations. Relative to their size, our human bodies are vast ancient continents. Invaders of a new land, the bacteria populate preferred habitats: some encamp on the open plains of our bellies, others scale the vertiginous heights of our eyelashes; some prosper in the boggy wetlands of our armpits; while others settle in the caves of our noses and our ears. As days go by, battles occur. Microscopic armies of new species sweep in to conquer territories and, ultimately, produce new colonisers to once again set forth in search of far off lands, jumping to another human host when close contact occurs in what is known as horizontal transmission. For other bacterial species, the association with us is much closer and they remain with us not just throughout our lives but also throughout the lives of our children on to whom they pass. These vertically transmitted species have interacted with us for many human generations. With evolution, the lines between us and these species become quite blurred. To understand how vertical transmission of bacteria works, let's look at a few examples. Babies in their mothers' wombs are thought to be free of bacteria, but scientists have found that during the birth process bacterial species on the vaginal wall are passed onto the baby's skin. It turns out these bacterial species are beneficial to a baby's digestive system and remarkably they turn up at just the right time to be transferred to the baby and passed on to the next human generation.

This close relationship between bacteria and humans brings to my mind one that ant species share with other organisms, which I studied during my doctorate degree. Ants forage for food as a well-coordinated army and to fuel this foraging they gather a sweet substance secreted by sap-feeding aphids, known commonly as blackfly or greenfly. This sugary honeydew is a waste product to aphids but a useful food for ants and, to return the favour, the ants protect the aphids from predators, carrying them to new pastures to feed, effectively shepherding them like tiny cattle. When ants form a new colony, young ant queens fly forth to find a suitable site, and many carry their honeydew-providing partners in their mandibles to start a new flock upon their arrival. It always touched me that these ant species, who are mostly highly predatory, could act both so intelligently and so gently. If you scaled an ant up to human size, their mandibles would be like industrial bolt cutters, yet they manage to successfully carry these soft-bodied insects to begin the partnership anew with the next generation of ants. Such impressive feats of behaviour arise through evolution because ants benefit from the interaction, and so they evolve adaptations – the gentle ferrying of their 'cattle' to new colonies – which facilitate vertical transmission of the aphids.

Another example of how our close dependency on bacteria has evolved can be found in the human immune system. Our bodies face a constant battle to keep out pathogenic – disease-causing – organisms that would harm us. Enemy 'non-self' cells are efficiently destroyed by soldier-like immune cells in the bloodstream, but what if those non-self cells are not enemies? What if some actually benefit us? Fortunately, our body's defences have evolved to tolerate these species, and our normal immune response is suppressed: the immune cells are stood down, avoiding a 'friendly fire' situation. To achieve this intelligent discrimination between 'good' and 'bad' non-self cells, beneficial bacteria species are presented to immune cells during their 'training', so they learn which cells not to attack. In essence, the bacteria are being treated like human cells and the immune system is being taught 'these guys are OK, they're one of us'. Obviously, this requires a fairly sophisticated adaptive immune system. It may be that our strong dependency on beneficial bacteria, and the need to carry a microbiome of thousands of resident species in our guts, has been a primary driver for the evolution of such an adaptive immune system.

In 2007, a scientist called Margaret McFall-Ngai observed that invertebrate organisms don't have such sophisticated immune systems. They only have 'innate' immune systems which lack the ability to discriminate between 'good' and 'bad' non-self cells. It just so happens that invertebrates have a much simpler microbiome with just a handful of species as long-term residents in their guts. This led McFall-Ngai to assert that our strong dependency on bacteria, and the need to carry a microbiome of thousands of resident bacterial species in our guts, may be a primary reason for the evolution of our adaptive immune system.

The evolution of such complex mechanisms for maintaining the bacterial ecosystem in our bodies has occurred because there are important benefits imparted by these species. Bacterial species in our guts work to break down the food we eat, making vitamins more easily available, benefiting not just our nutritional balance, but also our energy levels and even our moods. Disruption of the microbiome is increasingly linked with a range of maladies, both physical and psychological, for example depression. To test the causality behind this association there is some fascinating research in this area involving swimming mice. Yes, you read it correctly, swimming mice. The duration of time a mouse is prepared to swim before drowning is used as a measure of their motivation to survive. Depressed mice give up the ghost earlier and stop swimming (sad, I know, and I sincerely hope they pull them out of the tank at this point). Non-depressed mice swim for much longer, reflecting their higher motivation for survival. However, if these mice are given antibiotics, which disrupt their microbiome, they also begin to show depressive tendencies and fail earlier in the forced swimming test. It seems the bacteria in our microbiome may be more fundamental to us than first thought.

Bacteria also benefit us through simply taking up space that other more harmful species might otherwise colonise. In the war for dominance between bacteria, territory is everything, and if invaders gain a foothold, it can allow them to spread over large regions of the body. This is why taking antibiotics unnecessarily can be dangerous: it clears our body of benign bacteria leaving a blank canvas for renegade 'wild-card' species to exploit. 'Friendly' bacteria on our skin can prevent the colo-nisation by species that cause skin diseases,[4] while the damage to our gut microbiome is increasingly implicated in a range of other diseases including irritable bowel syndrome, obesity, Crohn's disease, colorectal cancer and coeliac disease, to name but a few.[5] Many of these diseases are 'immune related', meaning they are linked to a disrupted immune system with excessive inflammation responses. It seems our close evolution with the microbiome has led us to depend upon it so much that when the key species are lost our body does not function successfully, almost like losing a leg, or an organ like the kidney.

An 'evolved dependency' on organisms we closely interact with was shown in the 1960s by a scientist called Kwang Jeon, who was studying single-celled organisms called amoeba. His amoeba cultures accidentally became infected with patho-genic bacteria. The bacteria killed many of the amoeba, but a few survived. To Jeon's surprise, months later the descendants of these surviving amoeba had the bacteria inside them. In a further twist, if Jeon applied antibiotics to kill the bacteria, the amoeba also died. In a short timeframe of just a few months, the amoeba had evolved to completely depend on the bacteria.

How and why does this happen? Nature likes efficiency, so if a 'symbiont' (a partner organism living in close association with another) is producing a useful substance, then the genes in the first organism that produce the substance are redundant and may be put to other use or lost completely.[6] This evolved co-dependency between species works both ways: what starts off as a casual relationship can end up as a close mutual dependency. This seems to be the case with our human microbiome. If we disrupt the interaction with certain species inside us, we suffer. Our current lifestyle of processed-food diets, the use of antibiotics and the disruption of our biological clocks, through nightshift work and less exposure to natural light, appears to be disturbing our microbiomes and causing negative consequences for our health. Fortunately, as we start to understand the intricate dependency on our microbiome, new techniques to maintain or restore the beneficial partnerships are being revealed, although the science is still at a very early stage. Some of the proposed methods are more pleasant than others: the most palatable is simply changing our diet to eat more probiotics (foods containing beneficial bacteria) and prebiotics (foods which stimulate the growth of beneficial bacteria). The least palatable (literally) is probably the 'faecal transplant'. Here, a disrupted microbiome is restored through the ingestion of faeces from someone with a healthy microbiome. Disgusting, you may shout, but this treatment turns out to be as effective as it is repellent. Another method, that helps babies who have limited microbiomes, is the 'vaginal swab'. In caesarean births, the transmission of beneficial bacteria to babies during the natural birth process is bypassed, leading to babies with impoverished microbiomes, so doctors transplant the mother's bacteria, through vaginal swabs, onto newborns to facilitate their colonisation.

These methods of microbiome manipulation may seem crude, but the science of restoring microbiomes is still in its infancy. In the future, we might be inoculated with specific formulations of bacteria tailored to our unique genome. Until then, it is perhaps best not to disrupt the ecosystems inside us in the first place. Diet is obviously key here. We might think of the food we eat simply as fuel, akin to feeding a boiler with gas to produce energy, but these findings suggest we would do better to think of our diet as more akin to 'gardening' the human ecosystem. The food we eat is critical to determining which bacteria flourish inside of us and, given their actions influence our metabolism, affecting everything from our energy levels to our emotions, perhaps we should be more conscientious gardeners.

There is one final example to consider. So far, we have discussed the microbiome – all the non-human organisms colonising our inner and outer body surfaces – but picture for a moment the insides of any one of the 7 trillion human cells that are not red blood cells.[7] A microscope and special stain will reveal how each of these cells contains many tiny 'organelles' (literally meaning tiny organs) called mitochondria. These are the power stations of our cells, producing the chemical energy essential for our bodies to function. It turns out that these mitochondria were originally derived from free living bacteria, and the same is true for chloroplasts in plant cells, their power stations for photosynthesis. Around two billion years ago an event occurred which caused these proto-mitochondrial bacteria to be ingested and become part of a eukaryotic cell – the type of cells which make up all animals and plants. Evolution cemented this close relationship, through the 'evolved dependency' we encountered earlier – the bacteria lost many genes they didn't need because functions were provided by the host cell and, likewise, the host cell came to depend on the bacteria nearly exclusively for energy production.[8] The two genomes – 'nuclear' and mitochondria – are now closely co-evolved, producing chemical messages to regulate one another.[9]

So, it appears we humans are truly a chimera – microorganisms have not only colonised every surface of our body, but they sit within every one of our cells. We are truly interconnected with them and to say they provide benefits now seems like an understatement, at least to the extent that one might say our heart, or any of our other essential organs, benefits us. The boundaries between 'self' and 'other' are blurred and we might even refer to humans as a type of superorganism, encompassing the *Demodex* mites, bacteria, fungi and viruses that colonise nearly every part of our bodies. However, sometimes the goals of the microorganisms and our own can diverge. As we will see in the next chapter, microorganisms can sometimes use these close associations to manipulate us from within.

4

Hacked: is it really you controlling your body?

It's 2043, and ten years have passed since the Mars Rover 2 mission returned to Earth bearing the first news of extra-terrestrial life. The fact that the discovery constituted only simple bacterial lifeforms was at first underwhelming, although with the prevalence of these organisms found on nearly every habitat on earth it was a result that was probably to be expected. As the investigation into the new alien lifeforms developed, the story became a lot more interesting – they were carbon-based life, just like here on earth, and they shared a similar genetic code based on nucleic acids. However, there was just one fundamental difference: whereas genes of creatures here on earth encode for aspects of their own bodies and behaviours, these extraterrestrial bacterial genes were able to 'reach out' and manipulate the bodies of other organisms – they could produce chemicals to manipulate the nervous systems and physiology of other creatures, to their own benefit.

The findings created a stir in the world of biology, although the media only really started getting excited when it was revealed the containment procedures of the secure bio-facility back on earth where they were stored had been breached. After an extreme storm caused damage to one of the laboratories, an unknown number of bacterial cultures escaped into the environment. As weeks passed, the initial panic subsided. There was scholarly reflection in the science-focused media as to whether these genes had the potential to be passed on to other organisms.

That question was resolved on 15 September 2042 when a veterinary surgeon 2,500 miles away from the original containment facility discovered extraterrestrial genes in bacteria on the skin of rats brought into her surgery. The transfer of alien genetic material into terrestrial species was thought to have occurred via 'bacteriophage' viruses, which commonly ferry genes between bacteria. The colonisation of the genetic code of all Earth's organisms – its pangenome – by genes capable of manipulating the bodies of others had begun . . .

In the short space of several months the global media was buzzing with stories of new cases where extraterrestrial genes had been found in a whole range of microorganisms from bacteria to fungi and even some rapidly reproducing 'higher' organisms like insects:

NEWSFLASH 10.01.2043: Alien 'manipulator' genes have been found in a parasitic wasp species which attacks spiders. The wasp glues its eggs to the spider's body and the larvae that emerge begin to suck the spider's blood. During this process, wasp manipulator genes apparently produce some kind of chemical factors which are transferred to the spider's bloodstream and interact with its nervous system, causing its web-making behaviour to drastically change. The spider takes down its standard web and replaces it with a few thick cables that suspend the wasp and hold it safe while it turns into a pupa.

NEWSFLASH 12.03.2043: A parasitic fungus which infects ants seems to have obtained extraterrestrial manipulator genes which allow it to enter the ant's brain and control its host's behaviour. Rather than foraging on the forest floor the ant climbs into an exposed spot high in the canopy; there, fungal cells multiply and their stalks extend out from the ant's body, raining down fungal spores onto the forest floor below. These infect other ants to begin the manipulator gene life cycle anew.

NEWSFLASH 21.04.2043: A parasitic 'jewel' wasp is using manipulator genes to drive its cockroach host towards its burrow. The wasp directs its stinger into the cockroach's brain, specifically into the area that controls movements. Manipulator genes in the wasp encode for chemicals which, when injected into the cockroach's head, mimic neurotransmitters in its brain and 'drug' the insect into a docile state incapable of escape. The wasp is then able to lead the cockroach like a dog on a leash into its burrow.

You might think these apparently imagined cases of 'manipulator' genes bizarre and fascinating, or perhaps they seem a little worrying, given the potential for manipulation of us humans too. However, many of you will already know, or have guessed, that these are not hypothetical cases of extraterrestrial genes arriving on Earth. In fact, all the cases documented here are real examples of behaviours that exist on the planet today.

Manipulator genes are already widespread in many of the creatures around you. In some cases, scientists have identified the exact individual genes and the pathways of action which cause the changes in behaviour in other organisms. For example, in 2011 Kelli Hoover, an entomologist at Pennsylvania State University, studied a type of virus called a baculovirus, which attacks insects such as caterpillars of the European gypsy moth. While infected moth caterpillars unsuspectingly feed, the baculovirus multiplies inside them and, when at full capacity, produces enzymes which dissolve caterpillar cells into a slimy treacle-like goo. Virus particles then drip down to infect new individuals on plant leaves below. Because being higher up would make infections of other caterpillars more likely, this is exactly how the virus influences the caterpillar behaviour – before collapsing into virus-transmitting goo, the caterpillar climbs high in the tree. This is the exact opposite of its normal tendency to retreat to the ground to hide during the day. Scientists had been at a loss as to how this occurred, until Hoover and her colleagues identified a gene in the virus that causes the caterpillar body to produce an enzyme which disrupts its own hormones.[1] When they knocked out this gene in the virus – removing it from the virus genome – it led to the virus-infected caterpillars climbing downwards as they normally would when uninfected, instead of upwards. This is probably the first clear demonstration of how genes in one organism evolved specifically to alter the behaviour of another organism, though we have known about the existence of these manipulator genes for some time.

In 1976 the evolutionary biologist Richard Dawkins wrote

a ground-breaking book called *The Selfish Gene*, which popularised understanding of what genes are and how they work. One of the chapters was devoted to something Dawkins called the 'extended phenotype'. His suggestion was that genes within an organism not only affect their own physiology and behaviour (their phenotype) but can also reach out and effect changes in the world outside the organism itself. These projections outwards may involve altering the basic physical environment, as genes in the aquatic caddis fly do when they cause the fly to collect small stone particles. Other genes in the caddis fly code for proteins to make a silk and bind these stones together into a net-like structure which is used both to collect food and as a shelter for the insect. The design of this shell structure (which varies in composition and shape depending on the particular caddis fly species) is encoded for in the fly's genome, and so one might view it almost as an extension of the caddis fly's body itself. Rather than being made of insect cells, it is made of materials gathered from the fly's environment, but its construction is encoded for by DNA just like the living parts of the fly's body. Many other examples of such 'extended phenotypes' exist, such as the dams of wood made by beavers in rivers to provide food and protection from predators.[2] Dawkins proposed that not only the non-living physical environment but, in some cases, the bodies of other organisms might be manipulated. This is particularly the case for parasites, whose close association with their host is ripe territory for the evolution of manipulator genes to create an extended phenotype. Genes capable of manipulating host behaviour and physiology are not just the stuff of science fiction, they are already at large in our world right now.

If this manipulation of other organism occurs widely, then what about us? Are we also susceptible to genes from other organisms 'reaching out' to affect us? The answer is undoubtedly yes. We have already read how we carry within our human ecosystem many non-human organisms. Some of these are mutually beneficial, some are benign hitchhikers, others, though, are able to manipulate us to their own ends.

'Rabies' is a word that carries a certain weight. The rapid onset of the disease, its frightening symptoms – including foaming at the mouth, fear of water, aggression or paralysis and then, frequently, death[3] – make rabies widely feared. The disease is caused by a virus that is able to colonise most warm-blooded species; only birds seem relatively immune. The large majority of cases in humans are caused by bites from infected dogs. The frequency of rabies transmission is increased because infected dogs become very aggressive, and in turn are more inclined to bite. It is likely that these changes in behaviour are caused by manipulator genes contained in the virus genome. Attributing disease symptoms to the extended phenotype of a disease-causing organism is quite tricky. One necessary condition is that symptoms must clearly benefit the disease, for example, by increasing transmission rates. That certainly seems to be the case for rabies – if a rabid dog bites you, it enables the virus to spread. However, without looking into which specific genes lead to the change in host behaviour, as scientists did in the case of the baculovirus and caterpillars, we cannot rule out that the effect of the disease is simply a side effect. Yet, it does seem that for many human diseases, the way host behaviour changes clearly helps to spread the disease. For example, rabies replicates in salivary glands and is transmitted through the host biting another animal, and so swallowing or dilution of saliva reduces transmission. It turns out, therefore, that infected hosts develop 'hydrophobia', meaning they panic when presented with water and are unable to drink. There is also an increase in aggressive behaviour – normally placid animals turn wild and voracious, making them much more likely to bite and transmit the disease. Is the virus responsible for these changes and causing them deliberately? It is quite possible this is an evolved strategy, because the rabies virus is known to make use of cellular transport mechanisms in the host allowing it to infect the host nervous system where human behaviours are controlled.[4] More work is needed to understand exactly how rabies influences our nervous system, but it is certainly a strong candidate for containing manipulator genes.

Another likely contender for the manipulation of human behaviours is the human immunodeficiency virus (HIV), the cause of acquired immunodeficiency syndrome (AIDS). All viruses are expert manipulators to some degree, because they don't have the capacity to reproduce themselves. Rather, they hijack the machinery inside host cells and force them to make copies of the virus. The HIV virus is particularly adept at this – it contains over 400 genes that interact with its host to produce an environment in which the virus can thrive. The immune system is normally efficient at clearing pathogen invaders (think how a cold only lasts a week or two because the viruses that cause symptoms are eventually identified and eliminated from the body), but the HIV virus is able to evade it by 'hiding' inside the immune cells, sometimes for many years. Instead of eliminating the virus, human immune cells are iron- ically turned into factories that produce more viral particles. The immune cells autodestruct in an attempt to eliminate the infection, but this has a negligible effect on the virus because it has already infected many other cells, yet it cripples the host's own immune response making it less efficient. New immune cells arriving to an infected area suffer the same fate; like sol- diers turning their rifles upon themselves, they autodestruct.[5]

Because these diseases are relatively rare, you might think you yourself are unlikely to be the subject of manipulation by parasites. Yet, there are some parasites with this ability that are very common across the human population, although they do tend to be more benign in their health impacts. Take *Toxo- plasma gondii*, for example, a protozoan (single-celled) parasite found in the brains of many humans – around 30–50 per cent of us have it inside our brains right now.[6] Cats are the primary hosts in which this parasite replicates, while rats are used as secondary hosts who harbour and transmit the parasite. *Toxo- plasma*, however, can also survive in humans. The parasite has a clever means of manipulation to get from rats into cats. As anyone with a cat knows they are good hunters, yet rats are cautious creatures and will frequently hide or escape from their feline antagonists. To overcome this, *Toxoplasma* has evolved ways to manipulate the nervous system of rats – infected rats lose their aversion to the smell of cat urine, becoming posi- tively attracted to it. *Toxoplasma*-infected rats are also less risk averse by being much more active, more likely to explore novel, exposed environments, and worse at remembering their routes home.[7] All this makes them more likely to be caught by cats. We are now just starting to understand how the parasite induces these behavioural changes – it interferes biochemically with gene expression in a region of the brain which controls fear responses, effectively making the rats brave but foolish.

What about humans? We are not the intended target for *Toxoplasma* manipulation, we are just a 'dead-end' host (since our bodies are not eaten by cats),[8] but because our brains are structured similarly, many humans (males at least) also start to find the smell of cat urine more attractive![9] Having a less than normal aversion to the smell of cat wee doesn't sound so bad, yet a number of more serious impacts associated with *Toxoplasma* infection of humans are unfortunately being revealed. It turns out that those of us with *Toxoplasma* in our brains are more likely to be involved in traffic accidents, to commit suicide and to develop schizophrenia.[10] This is clearly a case where manipulation of our brains by another organism is something to be taken seriously.

Another example of a widespread agent manipulating us for its own ends is the common cold virus, or rather the large set of viruses that cause cold-like symptoms. These viruses replicate in our upper respiratory tract and cause symptoms including sneezing, a runny nose and coughing. Is it just a coincidence that these symptoms are also highly effective transmission mechanisms? Dr Lydia Bourouiba, a fluid dynamics researcher from MIT, who turned her hand to epidemiology – the study of disease prevalence and transmission – has used high-speed cameras to film sneezing events. She found that during these violent explosions, droplets of mucus are fired outwards in large arced trajectories of up to one to two metres. Furthermore, a turbulent cloud of finer droplets drifts up to eight metres and is suspended in the air for around ten minutes.[11] These droplets are loaded with virus particles that are easily carried into air conditioning systems (perhaps that's why one always seems to catch a cold on long plane trips). It is an extremely efficient way for a cold virus to spread, and it might well have evolved specifically to drive these coughing and sneezing symptoms in its host. This is still speculation, but it seems very possible.

Organisms causing food poisoning, such as the *Salmonella* bacteria, certainly manipulate our bodies by hijacking our immune cells to escape detection. The bacteria multiply in our intestines, which provokes an immune response that would potentially clear them from the body, except they have evolved a protein shell which makes them resistant to ingestion by 'macrophages' – the white blood cells that normally protect us by engulfing and digesting foreign substances in the body. Instead, when safe inside these immune cells the *Salmonella* disrupt electrical signals in the gut, tricking the macrophages to move out of the intestine and into other organs like the liver and spleen, where *Salmonella* becomes safely harboured.[12]

It seems humans are a common target for manipulation by other organisms. We may like to think that we have full control over our bodies, but other organisms have devised clever ways to control us remotely for their own benefit. Some of these cases are rare – like the rabies disease – but others – like *Toxoplasma* and cold viruses – affect many of us frequently. Even when we don't show active symptoms, viruses can also hide inside us, dormant in latent modes of existence. Many succeed in integrating themselves into the innermost, most sacrosanct, parts of our bodies – the genetic material inside our cells. Retroviruses integrate themselves into the instruction manual we use to build and maintain our bodies – they convert themselves into DNA format and invade our human genome. When our cells next transcribe their DNA, rather than manufacturing proteins to service our body, viral proteins are made instead and assemble to become functioning viruses. When the host cell eventually divides, replicating its DNA into two copies, the virus DNA is also faithfully replicated and passed on to the new cells until it leaps from our bodies to infect other unsuspecting hosts. So, the manipulators are very deeply connected with us. In fact, up to 8 per cent of our DNA is derived from viruses.[13]

Even a lot of our 'own' DNA actually doesn't function for the benefit of our bodies. After sequencing the human genome, geneticists found that over 70 per cent of it doesn't seem to code for useful genes and so this was termed 'junk DNA'.[14] It turns out this DNA simply doesn't function for us, but some has its own function to replicate itself, jumping across the genomes of different species like islands in the sea. It may actually be harmful to us – the inserted mobile DNA elements cause mutations in our code which negatively affect our reproductive success or survival. This led John Brookfield, the author of a 2005 review on this topic, to conclude that these genomic components are indistinguishable from parasites.[15]

In popular science texts, our DNA code is often likened to a computer program, containing the instructions to build all the different specialised cells our body needs to function. If that's the case, then our computer code has been hacked by viruses. It also contains malware in the form of parasitic DNA elements and our software can be remotely controlled by other organisms. Like poor firewalls, our body's defence systems have failed to keep these interlopers out, and they freely manipulate us to their own ends. Therefore, it is an illusion to think that this human body is solely our own. Not only in terms of our body's composition, but also in the way our supposed autonomy is subverted by the many other organisms which use us for food, homes or simply as transport, we are a super-colony whose behaviour is governed by many.

5

You are a copy and paste
from the book of life

I was much struck by how entirely arbitrary and vague
is the distinction between species and varieties

Charles Darwin, *On the Origin of the Species*

We have begun the journey of exploring our bodies and discovered we cannot easily define any unique independent existence. We are composed of the same basic material as not only the rest of our planet, but the whole universe. The molecules of our bodies will move on to form parts of other objects and organisms, once their transient, ephemeral roles in our bodies are complete. Our bodies are 'open' systems, with porous boundaries through which matter and energy continually pass and very few cells remain with us through our entire lives. What's more, a small army of non-human organisms hitch a ride inside and upon our bodies, and in subtle ways change its functioning to suit themselves. Given the ever-changing nature of your body and the fuzzy boundaries between it and the rest of the world, what is it then that ensures the person facing you in the mirror remains recognisable as 'you' throughout your life, albeit with the signs of gradual ageing?

Perhaps the answer lies in the 'instruction manual' of your body – the genetic code that is inscribed in your DNA containing the plans to build your body structure. What exactly is this genetic code? DNA stands for 'deoxyribonucleic acid' – it is a long, chain-like molecule consisting of nucleotides, each containing one of four bases (thymine, adenine, guanine or cytosine).[1] The ordering of these bases along the DNA chain forms a code with instructions on how to build protein molecules. As mentioned earlier, how this process works is often compared with the functioning of computer code. All the complex procedures performed by your desktop computer or mobile phone, such as displaying text, running computer games or making live video calls, are coded as information in its most basic form – a binary code – which is a very long sequence of ones and zeros. It is fantastic to contemplate how the complexity of a Beethoven symphony stirring our emotions, or the face of a loved one on a video call, can ultimately be reduced to a very simple code.[2] Our DNA code is similar, although rather than binary there are four possible options for each position in the sequence, determined by which nucleotide base is present. Our genetic code is read in triplets of DNA base pairs called 'codons' that code for twenty different amino acids. These amino acids then form protein molecules, the three-dimensional structure of which is hugely important in determining their function.

When cells divide, the copying of genetic code is not as accurate as the cut and paste function of computers, though the fidelity is nevertheless impressive. Each time a human cell replicates, all of the roughly 3.2 billion DNA base pairs are faithfully copied. Occasional mistakes do get made, but these mutations occur at a rate of only one in every 30 million base pairs copied. This equates to there being around 100 base pair differences between one cell and its immediate predecessor.[3] These differences accumulate as cells divide throughout your body, so if you have been told that every cell shares the same genetic code, then that is not altogether true: our bodies are a chimera of many slightly different genetic codes, although the difference between each of them is very small. These minor differences tend to have little effect on functioning, due to the redundancy with which the genetic code is written – several different codons can code for the same amino acid. So, for example, a codon which mutates from TCT (thymine-cytosine-thymine) to TCA, TCC or TCG, would still code for the same amino acid called serine.

Most mutations that have deleterious effects, especially those apparent early in life before we reproduce, tend to be 'selected out' of the gene pool – they cause their bearer to have a lower chance of reproducing successfully and so are less likely to be passed on to future generations. The high fidelity of DNA replication, and the fact that harmful mutations are weeded out by natural selection, mean that each of us shares a remarkably similar genetic code with other humans and also with many other species. We share around 95.5 per cent of our DNA with other humans, but we also share an impressive 37 per cent of our genes with simple primitive organisms like bacteria.[4] So although it is very true that we are closely related to each other, we are also remarkably similar to the rest of life on earth.

Because of the high fidelity of DNA replication between generations, we could say that our DNA code is 'borrowed' from our ancestors. We too, if we have children, will pass it on to future generations in much the same form. As Richard Dawkins describes, we don't need fossils to peer back into history, because that history is woven into the fabric of contemporary plants and animals in their genetic codes.[5]

What a wonderful thought that our human bodies are just self-repairing shells of matter (scavenged from stars), and the information in our cells is much the same as that in our mother's and father's cells and their parents before them, with much of the exact same genetic information found in many other life forms, including simple bacteria!

Of course, we are a *unique combination* of information and not simply a homogenous mass with the rest of life on earth. It is certainly true that for sexually reproducing organisms the process of 'recombination' means that we get a randomly selected set of half of our genes from our mother and half from our father, and so we are not identical to either. But this process does not generate fundamentally new information;[6] only the mutation of genes does that. It is more like shuffling a pack of cards before dealing them – we may each get dealt a unique hand, but we are all of us – all of humanity – ultimately from the same pack. And the differences between individuals are pretty minor when you really think about it – she has blue eyes, while he has green; he has whiter skin, while she has darker; she is taller, while he is shorter. Yet, all these people probably have two legs, two kidneys, a brain, a heart, an appendix, a liver, a small colon, a large colon and so on. The shuffling of genes leads to many superficial changes (as well as determining the biological sex of the person), while the basic DNA program is remarkably conserved, as it has been for many thousands of generations. It is a shared thread running through generations that weaves the garments of our body; certainly not something we can claim as our very own and the source of a discrete identity.

The strong similarity of DNA between different individuals and between different species presents problems for another commonly used analogy of the genetic code – the 'tree of life'. In this analogy, species are imagined as the tips of branches of a tree, distinct from one another and from the central trunk, which represents early ancestors from which all species evolved. This perspective is much better than the originally conceived views that humans were somehow distinct from Nature. However, the vision of a tree of life is also somewhat misleading because the separateness of the 'tips' of the branches highlight the distinctiveness between species, albeit with historic links between them, rather than the remarkable similarities in their DNA – species are connected to each other not just by their shared ancestry, but because they actually share much of the *same* information in their genetic codes. We humans are connected to each other not merely because we are similar in our DNA, but because we are all built mostly using *exactly the same* information. Contrary to the information that designs our bodies defining our identity, we share most of that self-defining information with others.

The tree of life analogy is also partially incorrect because there are in fact interlinkages between the tips on the tree

– there is sharing of information between individuals of different species. During this 'horizontal gene transfer', DNA is shared between individual organisms from completely different species, as well as the normal transmission of genes from parent to offspring in the same species (vertical gene transfer). In bacteria, this is hugely common, where there are multiple mechanisms by which DNA can be transferred. These include transformation – the uptake of 'loose' DNA from the environment, transduction – transfer mediated by viruses inserting DNA into the bacterial genome, and conjugation – transfer between two bacterial cells through direct cell-to-cell contact. Its ubiquity among bacteria has led researchers to raise the idea of a 'pangenome', defined as the total set of genes that occur across all the different strains of a species. This tendency to commonly exchange genetic material within and across species means that bacteria can evolve rapidly to exploit new environments. The result is that they are the most widespread organisms on earth, occurring in hostile environments, from deep-sea thermal vents to polar ice sheets and rock layers kilometres below the earth's surface. Less conveniently for us humans, it also gives pathogenic bacteria an unhelpful ability to evolve resistance to new antibiotics, making diseases less responsive to medical treatment. The fluid sharing of genes between bacterial species means antibiotic resistance appears more rapidly in bacterial populations than might be expected. The likelihood of resistance arising in an individual bacterial strain through mutation alone is very low, meaning it would take a long time for antibiotic resistance to arise. In contrast, when bacteria draw genes from a common pool – the pangenome – the likelihood of obtaining resistance genes is much higher.

Although bacteria win the prize for ease of horizontal gene transfer between species, the process also occurs in more complex organisms. Genes are frequently passed between different species by virus vectors that integrate themselves into host DNA, inserting both their own and another species' genes. Recent advances in DNA sequencing technology are revealing genes that have 'jumped' across from very distantly related species. For example, some aphids produce a pigment which gives them a red coloration. The genes to synthesise these carotenoid pigments are usually only found in fungi, plants and microorganisms. Aphids seem to have had the genes transferred to them from fungi.[7] Genes are most often transferred when organisms are in close contact, such as when one lives inside another. Many insects and nematode worms carry an intracellular bacterium called *Wolbachia*, whose genes become incorporated into the host genome. A similar process seems to have occurred in our human cells. As described previously, early in the evolution of the eukaryotic cells – those destined to diversify to create all plant and animal life of the planet – small bacteria were incorporated inside larger cells. They formed small membrane-bound sub-units producing essential energy for the cell that we now call mitochondria. Of relevance here is how original genes from these mitochondria have ended up in our nuclear DNA.[8] These genes are likely to have been transferred by small membrane-bound vacuoles containing digestive enzymes that break down molecules and act like miniature recycling centres inside the cell.[9]

Viruses are another mechanism by which genes are horizontally transferred between humans and other species. In some cases, we take the genes directly from the viruses themselves; for example, key proteins used in the development of our mammalian placentas are borrowed from viruses that first evolved the protein to fuse host cells together.[10] Over our evolutionary past, this process of gene-harvesting from across the tree of life seems to have occurred many times. It has been suggested that around 145 of our 20,000 genes have arisen from such horizontal gene transfer.[11]

The process also works the other way – with genes transferred from humans into other species. The protozoan parasite *Plasmodium vivax*, which causes malaria, is known to have obtained many genes from humans. It actually uses these genes against us to help it evade our immune system. In the future, with new genetic engineering methods, transferring genes between humans, and other organisms, could become even more common.

So, it seems clear that the borders between biological species are much more porous than we may have first thought, and certainly more than the tree of life analogy, with its linear progression of evolution from ancestral to modern forms, suggests. As the microbiologist Carl Woese writes: 'In questioning the doctrine of common descent, one necessarily questions the universal phylogenetic tree. That compelling tree image resides deep in our representation of biology. But the tree is no more than a graphical device; it is not some a priori form that nature imposes upon the evolutionary process.' Woese explains there is no real tree of life. Given the frequency of horizontal gene transfer, evolutionary descent does not operate in the solely linear Darwinian sense that we normally understand – it is essentially more network-like.[12]

This idea of the evolution of life as a dynamic interconnected network also features strongly in the work of Lynn Margulis, the founder of 'endosymbiotic theory'[13] and a strong proponent of the Gaia hypothesis with James Lovelock. Margulis suggested that evolution is not a linear family tree, instead it represents change in a 'single multidimensional being that has grown to cover the entire surface of Earth'.[14] Here, we see an extreme extension of the pangenome concept, encompassing all species on earth, past and current.

We should not lightly disregard the views of giants of evolution theory such as Woese and Margulis. It appears we need to radically revise our concept of species radiating outwards like branches in a tree of life, with the flow of evolution only in one direction. Instead, with DNA information frequently exchanged horizontally, there is strong interconnectedness between species comprising many evolutionary feedback loops in our complex, tangled web of life. With this in mind, perhaps we should refine the analogy of DNA. Instead of viewing each organism as having a unique computer program inside it, we now know copies of much the *same* information exist in many different biological machines. The information can be transferred between biological machines, as when a whole program is copied to create a new unit (vertical transmission of information during cell replication). Smaller units of information can also be passed directly between machines (during horizontal gene transfer), as we do when using a data pen to transfer useful computer files. Life is therefore more like a great networked cloud computing drive, where information is used to build bodies scavenged from material in the environment. So take a look in the mirror right now. The architect of your body is your DNA, the amazing coded information packaged in each of your cells. But whose DNA is it really?

PART TWO

OUR INTERCONNECTED MINDS

We are like islands in the sea, or like trees in the forest. The maple and the pine may whisper to each other with their leaves.... But the trees also commingle their roots in the darkness underground, and the islands also hang together through the ocean's bottom. Just so there is a continuum of cosmic consciousness, against which our individuality builds but accidental fences, and into which our several minds plunge as into a mother-sea or reservoir.   William James, *Essays and Lectures*

No man is an island entire of itself; every man is a piece of the continent, a part of the main   John Donne

Devotions Upon Emergent Occasions and
Seuerall Steps in my Sicknes – Meditation XVII

6

All your ideas are crowd-sourced

The brain that is interpreting these words is a remarkable thing. It's packed with around 170 billion cells, mostly neurons, which in turn are packed into an intricately folded layer 3–4mm thick called the cortex. The dense folding of the cortex increases its surface area, allowing neurons to connect to each other using their exceptionally fine projections of the cell membrane, which look a bit like the ultra-fine roots of a plant. The finest of these wire-like connecting filaments, called neurites, are around 0.1 micrometre thick. Ten thousand of them side by side would be just one millimetre wide, and the total length of them in a human brain sums to millions of miles, facilitating a staggeringly complex level of connectivity. These connections are how our brains store information, in the form of waves of electrical activity coursing through networks of neurons. Each neuron has the potential to connect to thousands of others at the same time, meaning that the number of possible ways the human brain can be wired up is greater than the number of atoms in the universe. This potential overall connectivity in the brain is like a blank canvas on which, in any person, at any given time, only a subset of neurons are actually connected – referred to as the 'connectome'. This connectome delimits our thoughts, personalities and memories – all are contingent on the way our neurons are wired up. Under a materialist view of the way our minds function (the only one supported by science to date), our conscious experience depends on the configuration of neurons and the waves of electrical activation flowing through them. This neural activity in the brain is closely linked to the rest of our body through nerve impulses and chemical messages flowing in both directions, as well as to the outside world through our senses.

It is strange to think how, if we could somehow modify the connections between the neurons in our brains, we could share the exact thought processes and personalities of any other individual in the world, even geniuses like Albert Einstein, Ada Lovelace, Isaac Newton or Marie Curie. In fact, we can go further than that. In terms of the potential network of connections between neurons, any thought that has ever occurred or ever will occur in a human mind is, theoretically, already in our head, it just needs those connections to be made physically rather than potentially. You hold infinity (almost) in your head! But there's the rub; *how* our brains develop and connect up those 170 billion neurons is crucially important to how we think and who we are.

As a human child develops and learns, connections are made between neurons. Initially this occurs at an astounding rate. Experiments in mammals such as rats show up to 250,000 new connections are made every second.[1] Each of these connections involves the creation of millions of gateways between neurons

(termed 'synapses'). The density of synaptic connections between cells is thought to be around 1,000,000,000 synapses
per mm³. Across these synapses, in response to electrical stimulation, neurotransmitter chemicals (such as dopamine and
serotonin) are fired outwards from one cell to receptors in the
surface of the receiving cell. The build-up of neurotransmitter
chemicals in the receiving cell leads to electrical activation
and the propagation of the electric signal. So although it is
tempting to think of our brains just like computers working
by electricity, chemical transmission of information is a key
part. Each time you have a thought in your head, there are tiny,
immensely rapid surges of chemicals linking up the neuronal
networks.[2] At the same time as new synaptic connections
are being formed between brain cells, others are being lost.
It is similar to a huge rail network responding to demand by
building new intersections while decommissioning others.
This apparent merry-go-round of activity is how our brains
learn: new patterns of activation are cemented by experience
with synaptic connections enhanced in a process discovered
by neuropsychologist Donald Hebbs and summarised as 'neurons that fire together wire together'.[3] Neuronal connections
frequently used become like well-trodden roads that are more
easily traversed, hence the improvement of skills with practice.
Neuronal connections which are used less, on the other hand,
become like overgrown tracks, narrowing so much they eventually disappear. The variation in connectivity between brains
explains the differences in personalities and skill sets between
people, while the dynamic nature of our connectome explains
changes in personalities over time. We differ from earlier

versions of ourselves – we are not the same person we were five minutes ago, let alone ten years ago.

The overall set of connections between our neurons – our connectome – is responsible for our current state of consciousness and sense of identity. It modulates incoming information from our senses, affecting how we perceive, interpret and react to the world. Let's think about the process of perception for a moment. Look at an object close to you now, examine it in detail – the colours, the reflection of light upon it, the shadows in the textures. Now, consider how what you are perceiving is not 'out there' in front of you, but is actually just a representation of the object inside your head. The real object 'out there' is reflected in the shimmering activation of millions of neurons across your brain. Turn your head slightly to pick out another object, and those waves of activation rapidly reconfigure within microseconds to reflect that object instead. It is tempting to think that visual perception works like a camera, with our eyes like the lens transmitting the full high-resolution image directly to be interpreted by our brains. However, receiving so much information from the senses, continually updated every microsecond, would overload our brains and reduce their ability to detect useful signals relevant to our survival and reproduction among a mass of 'noise'. Instead, our attention is selective and directed towards aspects of our surroundings we predict to be relevant. This is exemplified by the near-famous psychology experiment in which participants are asked to watch a basketball game paying particular attention to something specific, such as the number of times players in white shirts pass the ball.[4] Halfway through the video a man dressed as a gorilla walks right through the middle of the game. Surely, you would notice something so apparent and unusual? Yet half of the participants when asked if they saw a gorilla had no idea what the experimenter was talking about. Experiments such as these reveal how blind we are to things we are not specifically directing our attention towards.

How and what we pay attention to depends on the way our brains are currently configured – the state of our connectome. This itself varies over time, depending on the extent to which we are in an outward-looking versus introspective mood, and as a result of our expectations at any given moment. The focus of our attention also differs greatly between people, because although our individual connectome varies constantly, outweighing this internal variation are the substantial and consistent differences in the connectomes between us, which underpin differences in personality. It's not surprising that someone with a great love for the natural world is always pointing out plants and animals, while someone with a passion for locomotive trains notices the subtle differences between engine carriages. They can both look at the same picture of a rural train station set among trees and hedges and 'see' different things. In addition to our different filters for perceiving the world, the configuration of our neural networks determines how we *interpret and evaluate* any given input.[5] You probably have friends, for example, who seem to repeatedly find positive aspects in some situations, while others dwell gloomily on the negatives.

With all these differences in our connectomes, it's surprising we don't seem alien to each other. There is a famous psychology book that was popular in the 1990s entitled *Men Are from Mars, Women Are from Venus.* Beyond broad differences between the sexes, however, on the basis of how our brains idiosyncratically perceive and interpret the world, one could justifiably conclude that every individual is their own planet, distant and isolated from others. In his novel, *A Tale of Two Cities,* the author Charles Dickens reflected on this idea:

> ... that every human creature is constituted to be that profound secret and mystery to every other. A solemn consideration, when I enter a great city by night, that every one of those darkly clustered houses encloses its own secret; that every room in every one of them encloses its own secret; that every beating heart in the hundreds of thousands of breasts there, is, in some of its imaginings, a secret to the heart nearest it!

Dickens proposes this is a 'wonderful fact', although I find the idea that we are all isolated and secret to each other rather dark and gloomy. So where does that leave the central idea of this book – that we humans are more intimately connected with each other than we intuitively think? It is correct that our minds are all different, and it is indeed wonderful how every one of the billions of people on the planet is unique. But, just like our DNA code, being unique does not mean that we are independent. Our minds are dynamic and ever-changing; they are gems unique to us, but they are also intersections to the world with information flowing inwards from all directions. A huge source of information entering these busy crossroads comes from other people. When we interact with one another, almost like a science-fiction movie, the patterns of neural activity that represent concepts in each of our minds can be transferred between our connectomes. Unlike the horizontal transfer of DNA information between individuals, the horizontal transfer of information between our connectomes is not a rare event. Rather, on a near continual basis, information flows between connectomes, so much so that our brains are more like open doors than sealed treasure chests.

Let's try a thought experiment before we explore some of the mechanisms behind this information transfer. Think of a green triangle with a pink circle in the centre. Hold that image in your head. Got it? OK, we have just successfully achieved remote transfer of a 'neuronally represented' concept – a concept reflected by information stored in electrical currents pulsing around the neuronal network in your brain. As you visualised the triangle and the circle, the set of neurons fired in your brain to represent these shapes, just as they fired in mine when I wrote this sentence. The spatial layout of those neuronal networks in your brain is likely to be different from mine, because we are each 'wired up' slightly differently; but the concept triggered in your mind was equivalent.[6] So a concept has jumped easily between two, apparently discrete, minds, not only across a potentially great distance (depending on where you are reading this) but also across time. That is the magic of the written word. Those vast libraries around the world are not just filled with dusty books, they contain perfectly stored information from the connectomes of millions of people throughout the ages. When those authors put pen to paper the concepts represented by neurons in their brains were translated onto the page, just waiting to leap from the page into the neurons in our heads when we read the text years later.

In the internet age, the transfer of information between minds has become so much easier. With the capacity for near-instant communication, the potential to accelerate the accumulation of human knowledge and understanding is increased. Yet some are sceptical about this. Two American mathematician brothers, Donald and Stuart Geman, published a letter in 2016 in the well-reputed scientific journal *Proceedings of the National Academy of Sciences,* suggesting technological improvements in communication may be responsible for a (perceived) lack of great scientific innovations in the last 50 years.[7] They attributed this to too much communication leading to 'group think' and less independent creativity. The brothers made an analogy between the evolutionary process that leads to diversity – where species need some degree of in-dependent isolation to evolve – and the diversification of new ideas. It is true that complete homogenisation of the gene pool between species prevents the formation of new species, yet the fact is that *some* horizontal transfer of genes between species does still occur, and we still see the evolution of new species. Similarly, communication between people can still allow them to develop original ideas, and, what's more, it may even accel-erate new insights. Just as the bacteria we encountered earlier evolve more rapid solutions to antibiotic chemicals when they share genes, so too do we innovate faster when we share ideas. A classic example are the letters between Charles Darwin and Alfred Wallace that clearly show how sharing information helped them refine the theory of evolution. Darwin often receives the credit, but Wallace and many others, like the geneticist Gregor Mendel, provided substantial contributions that the theory of evolution was built upon. Similarly, the discovery of the helix structure of DNA was a strong team effort between Rosalind Franklin, Maurice Wilkins, James Watson and Francis Crick (though the former two often get less recognition). In the modern world, with the capacity for near-instant international cooperation, these scientific teams are much bigger, allowing the development of huge innovative projects where each scientist provides a small jigsaw-like piece to the overall solution. Examples of this include: the Human Genome Project – helping us to understanding our entire DNA code; the Large Hadron Collider – designed to unlock the fundamental physics of the universe; Neptune – the world's largest undersea observatory; the Advanced Light Source – a particle accelerator creating the ultimate microscope, and the European Extremely Large Telescope – which does exactly what it says on the tin.

We like to give credit to single lone geniuses, yet the history of science shows that great ideas are very rarely the product of a single mind. Rather, they are the culmination of knowledge produced by many minds over history. It may be a hackneyed saying, but each of us has an opportunity to stand on the shoulders of giants and push the frontiers of human knowledge outwards. Evidence of this interconnection is the frequency with which the same idea or innovation whose 'time has come' can appear almost coincidentally in different people. The incandescent lightbulb, thermometer, telephone, steamboat and hypodermic needle were all invented by multiple people in different locations, though they tend to be attributed historically to a single inventor.[8] Although these innovators may have appeared to converge upon an idea independently, they were synthesising knowledge from a common pool. If you want to see this for yourself, have a look at the *Edge Question Series* edited by John Brockman. In these stimulating books, a single question is asked of many well-known thinkers, authors, philosophers and scientists. Brockman organises the responses by disciplinary background of the author, and it is fascinating to see how many individuals seem to be converging on the same new ideas. It seems by connecting up our connectomes, through the written and spoken word, we rapidly advance the sum of human knowledge. Creation and innovation is not the domain of geniuses working independently, but an outcome of the whole interconnected human endeavour.

Beyond the spoken and written word, neuro- and cognitive-scientists have made further advances over recent years in our understanding of how our connectomes interact with those of other people. One particularly important, more primitive, process is called 'resonance'. Here, the neuronal patterns triggered are not conceptual thoughts, but emotional states. When we see a video of an innocent child crying, we may enter a state of empathy and feel a deep tug of emotion reflecting the pain that they are experiencing. It turns out this emotion is not just *like* the pain we are observing, it is the exact same pain centres in our brain that are activated, suggesting that in this state of empathy we really do feel the pain of others.[9] The resonance process works equally for joy as well. Watch someone smiling and laughing and you instinctively feel those expressions rising within yourself too.

A related process is 'mirroring', where we subconsciously mimic the actions of others through mirror neurons being activated in relation to both observing and carrying out the same action. So when the person you are talking to has their arms folded and is leaning against a post just as you are, the mirroring system in their connectome is probably active (and likewise in yours). There are several explanations for the function of this mirroring and the argument has not yet been fully resolved. It may exist to help us understand the actions and intentions of others better, or it may improve social bonding and cohesion – establishing rapport, or perhaps it evolved to help us mimic and learn new skills from others. Whatever the answer, the behaviour is widespread in humans and other primates.

Charles Dickens was also deeply aware of this phenomenon and the transmission of emotions between people. In *A Tale of Two Cities* he describes a courtroom trial:

Any strongly marked expression of face on the part of a chief actor in a scene of great interest to whom many eyes are directed, will be unconsciously imitated by the spectators. Her forehead was painfully anxious and intent as she gave this evidence, and, in the pauses when she stopped for the Judge to write it down, watched its effect upon the counsel for and against. Among the lookers-on there was the same expression in all quarters of the court; insomuch, that a great majority of the foreheads there, might have been mirrors reflecting the witness.

This is somewhat contradictory to the earlier suggestion by Dickens, in the same novel, that every heart is a locked storeroom. In contrast, emotions seem able to transfer swiftly between people and, if our hearts are locked, others may well hold the keys. Or perhaps our hearts are not locked away at all, but we 'wear them on our sleeves', as the expression goes. Beyond platitudes, the idea that there are strong and consistent connections through the transfer of neuronal patterns between people's minds at both the conceptual and emotional level aligns with the scientific evidence. Of course, we cannot necessarily transfer our whole connectome, to really 'see' the world from someone else's perspective. Perhaps in the future we may be able to 'download' someone's connectome by transferring it to a computer, as speculated by Sebastian Seung in his book *Connectome*, but that's unlikely to happen for a while (and don't hold your breath, given the sheer complexity of the brain).[10] So we will have to make do for now with partial transfer of neuronal patterns in the connectome through written and spoken word, music and art, and processes like resonance and mirroring. These tools, used artfully, can give us an excellent window into the workings of other people's minds. Also, as many people will know, spending many years together with someone you love gives a deep empathetic intuition about how their connectome works, albeit we will never see the world completely through their eyes.

Perhaps Dickens is right in his gloomy prognosis that we can never truly fully know the secrets of another's heart. Yet, it is clear that our minds are far from being isolated islands. We are continually sharing information between connectomes through written and spoken word, as well as empathy processes like resonance and mirroring. And there are other non-verbal communication methods we haven't touched upon, such as music and art. Of course, we each interpret these media in our own ways depending on the current configuration of our connectomes, as we do any other inputs, but there seem to be strong commonalities in our emotional responses. A musician can write a music score and intentionally influence listeners to feel triumphant, thoughtful or sad. Imagine an orchestra playing a Bach composition to an audience of a hundred people. With slight variations, the same general pattern, reflecting the musical notes, is shimmering in the neurons of everyone's head during the concert. What magic, that complex neuronal patterns in the head of a German man over 250 years ago can be translated to a musical score and then into vibrations in the air, which resonate in the eardrums of hundreds of people causing the initiation of neuronal patterns, which to a strong degree match the original information and perhaps the same emotions in the head of Johann Sebastian Bach.

Emotions are also transferred between people by chemical messages in the air. We cannot detect these consciously, yet we have our own individual pheromone signature, which also varies depending on our mood. Our pheromone scents are picked up by others and lead to the transference of emotional states. A neat experiment by researcher Lilianne Mujica-Parodi and colleagues in 2009 took sweat samples from the armpits of novice skydivers and aerosolised them for unknowing participants to then breathe in.[11] As a control, a second set of participants breathed in odours produced from the armpits of individuals running on a treadmill (producing the same amount of sweat but without the adrenaline). The two types of treatment were indistinguishable in terms of the smell – both mildly sweaty. However, under a brain scanner, the odour from the skydiver group, but not the treadmill group, elicited heightened responses in the amygdala brain region of participants, a neural region associated with fear responses. Mujica-Parodi suggests there is a hidden biological component to human social dynamics. Emotional stress seems to create an alarm pheromone that causes contagious emotional responses in others. Incidentally, a later study by a different team in 2017 found that these stress-induced pheromones can cause reduced performance in others exposed to them. Dental students performed less well in a tricky task when mannequins they were performing dental treatments on were dressed in T-shirts worn previously by students in a stressful exam (versus T-shirts worn by students in a calm lecture, as a control).[12] So, here is a life lesson: don't go to the dentist after you have been skydiving or have done a stressful exam.

We are connected to other people's connectomes for the most part of every day. You might be reading a book such as this one and connecting through the written word, while at the same time hearing sounds of voices or music in the background or being aware of someone else in the room, linking your neuronal patterns to them at the same time. It is hard to suggest that we are ever truly isolated. It really is just a matter of perspective – which, fortunately, we can shift because our connectomes are dynamic.

7

Love thy neighbour
(then copy them)

In winter, the temperature in Manitoba, Canada, falls as low as −40°C overnight and can remain below −18°C for weeks on end. Small isolated townships are hemmed in by thick blankets of snow. There is one such town, comprised of First Nation indigenous people, in which a very unusual series of events once occurred. The township is accessible only by air, or during a narrow six- to eight-week window in mid-winter when a network of roads is created across the frozen province, temporarily linking people to the rest of the inhabited world. Perhaps this isolation and the inhospitable climate go some way to explain why a series of unhappy events unfolded in this specific location.

It all started on 1 February 1995, during the final of the four darkest and coldest months of the year. A fourteen-year-old-boy was found dead after hanging himself.[1] Exactly one month later, a sixteen-year-old-girl was found dead in a similar fashion. The suicidal deaths of these two young people in a small community of less than 1,500 people was terrible, but they were unfortunately to be the first of many. In the subsequent month there were two more suicides: two males in their early twenties, one killed by hanging, the other by shooting himself. Three other males aged between twelve and fifteen also tried to hang themselves but were unsuccessful. The next month, April, saw three further hanging attempts and one drug over-dose. As the chance of these events all arising independently in such a small sample size is infinitesimal, the conclusion has to be that they were in some way connected. People referred to the phenomenon as suicide contagion – and it wasn't over yet.

Things accelerated in May. Four girls tried to kill themselves with drug overdoses, then an eighteen-year-old boy and a seventeen-year-old girl tried to hang themselves and a twenty-three-year-old boy shot himself (non-fatally). Four other children were airlifted out of the town, on the basis of psychiatric assessment suggesting they were also at high risk of attempted suicide. Despite these efforts, the authorities were unable to prevent the suicides of a fifteen-year-old girl and a twenty-one-year-old boy who were found dead from hanging themselves.

What was the cause of this terrible tragedy, which ulti-mately led to six deaths and tens of attempted suicides in just a few weeks? It is known that when suicides are reported in the media, depending on how they are communicated, it can affect the frequency of further suicide attempts. Most news reporters are now trained to cover suicide in a particular way to prevent this. At the most basic level, recent suicides prompt people to consider the possibility of suicide themselves. How-ever, beyond this, in the small Manitoba town it was clear that some of the suicide victims interacted with each other. At least three of them were close friends, and two had written notes to each other about the possibility of ending their lives. These

social interactions suggest that suicide can be a 'contagious' behaviour, and in groups of friends in which a suicide occurs the
risk of further suicides is elevated. Some might question the
idea that the behaviour is genuinely contagious, and point to
the fact people make friends on the basis of similar personality
traits (hence the saying 'birds of a feather flock together'), so
it is expected that traits like depression, which might lead to
suicidal tendencies, will be clustered among groups of friends.
This is likely to be a contributing factor. However, there is
also evidence of suicide contagion among people who have
not self-selected into friendship groups – work colleagues, for
example. A study of 1.2 million people in Stockholm, Sweden,
found that workers who knew a colleague who committed suicide were 3.5 times more likely to commit suicide themselves.[2]

Studies of social networks have found that many other
tendencies and behaviours seem to be transmitted readily between people. These include things like obesity risk, taste in
music and voting choices. People who are closely connected
in social networks commonly share behaviours and opinions,
even after accounting for self-selecting into similar groups.

Of course, we have known about the potential for other
people to influence us for a long time. The Dutch philosopher
Spinoza wrote about how our decisions are strongly influenced
by other people, even though we feel like we have completely
free will to make up our minds. However, with the rise of
the internet and social media linking us up as never before,
some researchers are sounding warning notes. According to
psychologist Robert Epstein: 'There's something happening
here which is really unprecedented. Technologies are rapidly

evolving that can impact people's behaviours, opinions, attitudes, beliefs on a massive scale – without their awareness.'[3]

Historian Yuval Noah Harari describes how scholars increasingly see culture as a kind of 'mental infection or parasite where humans are the unwitting host'.[4] In 2016, world politics became increasingly polarised, with views from the fringes taking centre stage. Britain voted to leave the European Union, a hugely costly endeavour, and America voted in President Donald Trump. Many pundits have pointed to the media, both in print and online, transmitting messages with very little basis in fact, as having a strong effect in changing people's voting tendencies. Another clear example of harmful 'memes' proliferating through the internet is the spread of anti-vaccination campaigns. Scepticism about the safety of vaccination is not new, it has been around since the practice first began (the Chinese Emperor K'ang Hsi supported inoculation of smallpox after his father died from the disease in 1661, and he wrote about having to overcome negative public opinion to ensure widespread adoption).[5] Of course, being sceptical can be a good trait, especially in light of limited facts, yet modern science is now absolutely clear that the danger of severe side-effects from vaccinations is often vanishingly small, and in some cases widely purported links, for example between the MMR – measles, mumps and rubella – vaccine and autism, are completely non-existent. However, social media and the internet have allowed conspiracy theories to thrive, with people seeking out spurious information that confirms their viewpoint, while interacting with like-minded others in echo chambers where these views are shared and reinforced.[6]

As a consequence, there has been a recent spike in measles outbreaks across the world. The World Health Organisation reported a 30 per cent increase in reported cases between 2016 and 2017, causing an estimated 110,000 deaths globally from the disease in 2017. In many places, like the European Region, this trend has accelerated further still, from just over 5,000 cases in 2016, to over 82,500 cases in 2018.[7] This has happened because changing attitudes have led to the proportion of people vaccinated to fall below the 95 per cent threshold needed to allow herd immunity.

So it seems that evidence is accumulating on how ideas are rapidly propagated from one mind to another across social networks, and there are growing concerns about how these can influence each of us every day to behave more in line with the opinions of others.[8] In the face of this growing phenomenon, the World Economic Forum has listed 'digital wildfires and misinformation' as among the most significant geopolitical risks we face today.

Of course, on the plus side, not everything that spreads through social networks is bad. Intrinsically, the porous structure of these networks that connect people is neither positive nor negative, they are simply neutral mediums. Because of our strong social tendencies as primates, our connectomes can be very effectively joined up to form 'super highways' between our minds. We can actively choose to use these interconnections as positive forces for good, in order to counter misinformation in the world. This is especially necessary with new technology causing rapid expansion of social networks. Positive action may be needed to ensure they develop in a way which benefits society overall. As Einstein once said, 'The world is a danger-ous place, not because of people who are evil, but because of the people who don't do anything about it.' Here are a few examples of people who are doing something about it.

The website PatientsLikeMe[9] joins up nearly half a mil-lion people, many with rare and unusual medical conditions. Through an online network they share information on what state-of-the art treatments are available and seem to work best for them, and support each other through simply being able to share their experiences with someone who understands better what they are going through. Without this online network, the chances of meeting someone with a similar rare condition in your local area would be pretty slim. In other cases, people are harnessing connectivity to benefit those who cannot help them-selves. The International Network of Crisis Mappers[10] – actually a meta-community of organisations – combines geographic information systems, mobile technology and crowdsourcing to map humanitarian crises in real time. These systems provide essential information allowing disaster response teams such as medics, firefighters and aid workers to be rapidly directed to the most important locations. They have proven themselves in cases such as the 2010 Haiti earthquake and many times since, leading to thousands of lives being saved. A final example is the Atlas of Environmental Justice,[11] a publicly available live dataset of envi-ronmental conflict locations around the world, serving to raise awareness, provide a platform for networking and a resource for advocacy. There are a growing number of environmental activists kidnapped or imprisoned around the world and this tool raises the visibility of such cases, allowing a critical mass of social and political pressure for their release. In other cases, it flags areas where the land and water of local communities is contaminated by mining or chemical effluents from heavy industry, or where violence is used to illegally force people from their homes to carry out environmentally damaging activities.

These examples demonstrate how connecting people through new technologies can lead to huge mutual benefits. Some people are scared about the pace of technological development, and many, including well-regarded minds such as Bill Gates and the late Stephen Hawking, have voiced concerns about the unbridled expansion of information technology and artificial intelligence. Yet there is no going back to a world that is less connected; extended social networks facilitated by technology are here to stay. As we have learned, we humans are social animals with brains wired to connect to others, allowing the fluid transfer of thoughts and emotions across social networks. In the twenty-first century, new technology has allowed us to expand these networks, and so our power and reach to influence others has dramatically increased. We now need wisdom to guide the development of these technologies so they function for the benefit of society. Our human identity does not end at the boundary of our bodies and our minds are deeply connected to one another, which makes the curation of these social networks, and society in general, a simple and logical act of self-care. This requires developing institutions that constrain the use of social networks for negative purposes, such as promoting terrorism, misinformation and xenophobia, and instead allow the more positive and progressive connectivity between billions of humans across the globe to flourish.

8

You cannot survive as a separate unit

It really boils down to this: that all life is interrelated.
We are all caught in an inescapable network of mutu-
ality, tied into a single garment of destiny. Whatever
affects one destiny, affects all indirectly

Martin Luther King Jr[1]

We humans are the single most mutualistic species on the
planet, capable of working together in intricate and wonderful
ways. Pick up a pencil. Such a simple and seemingly obvious
object. It is just a stick of wood covered with lacquer, surrounding a graphite lead core, with a bit of metal and rubber stuck
on the top. But could you make a pencil exactly like that yourself? Just think for a moment what it would entail. In 1958, an
American businessman, Leonard Read, did so and wrote down
his thoughts in a short essay called 'I, Pencil'.[2] To begin, Read
points out that making a pencil is not at all simple:

> Simple? Yet, not a single person on the face of this earth
> knows how to make me. My family tree begins with what in
> fact is a tree, a cedar of straight grain that grows in Northern

California and Oregon. Now contemplate all the saws and trucks and rope and the countless other gear used in harvesting and carting the cedar logs to the railroad siding. Think of all the persons and the numberless skills that went into their fabrication: the mining of ore, the making of steel and its refinement into saws, axes, motors; the growing of hemp and bringing it through all the stages to heavy and strong rope; the logging camps with their beds and mess halls, the cookery and the raising of all the foods. Why, untold thousands of persons had a hand in every cup of coffee the loggers drink!

The logs are shipped to a mill in San Leandro, California. Can you imagine the individuals who make flat cars and rails and railroad engines and who construct and install the communication systems incidental thereto? These legions are among my antecedents.

I think this last sentence is wonderful. Using the word 'antecedents' (things that logically precede another), Read emphasises how the production of the pencil depends on so many other people and their inventions. He then goes on to describe the intense, harmonious cooperation between humans required to bring the pencil into existence – the millwork, the transportation, the mining of the graphite; activities not restricted to one continent but involving thousands of people across the globe. Everything, right down to the seemingly simple lacquer, is the work of many hands and minds:

Do you know all the ingredients of lacquer? Who would think that the growers of castor beans and the refiners of castor oil are a part of it? They are. Why, even the processes by which the lacquer is made a beautiful yellow involve the skills of more persons than one can enumerate!

Given all this knowledge from thousands of minds, Read claims that no one person has the expertise to build a pencil. He ventures further, that all these tasks to manufacture the object are not the coordination of some great mastermind, but rather arise as an emergent property from the demand for pencils and the cooperative nature of human society. Note, this cooperation is not driven by some great altruism, but rather a simple outcome of the individual benefits that accrue to individuals by trading their wares and services. Economists often refer to this as the 'invisible hand':[3]

Actually, millions of human beings have had a hand in my creation, no one of whom even knows more than a very few of the others ... There isn't a single person in all these millions, including the president of the pencil company, who contributes more than a tiny, infinitesimal bit of know-how ...

I, Pencil, am a complex combination of miracles: a tree, zinc, copper, graphite, and so on. But to these miracles which manifest themselves in Nature an even more extraordinary miracle has been added: the configuration of creative human energies – millions of tiny know-hows configurating naturally and spontaneously in response to human necessity and desire and in the absence of any human master-minding!

Who would have thought a simple pencil had such a

complex history. And the same is true for every human-made object around you now, including the clothes you are wearing. Literally thousands of people had a hand in their design, manufacture and transportation. And supporting those people and their machines were armies of others. An immense and intricate web of interconnections reaches out from every object, linking up millions of people across the planet.

There is a Latin phrase *e pluribus unum* meaning 'from the many, one'. The words are borne as an inscription on American coins, referring to how people from many colonies once joined together to form one nation, yet it is an equally apt way to refer to the objects that humans produce. From the many people, one object – each single manufactured object originates from the craft of many thousands of people. We are highly dependent on others for pretty much everything in our lives. Our clothes, phones, shoe, this book – all draw on antecedents crafted by legions.

Incidentally, computer scientist Jon Kleinberg uses the term *e pluribus unum* in a slightly different way. He refers to how some 'things' that humans make, such as our Gmail or Facebook accounts, are not actually one discrete thing but 'distributed entities'. User interfaces give the illusion of a single entity, but Kleinberg points out how these actually comprise many physically dispersed components scattered across computer servers all over the world. These parts work 'independently but co-operatively to produce the illusion of a single unified experience'.[4] There are parallels here between how these physically dispersed computer programs operate and how the human mind works – another 'seemingly-unified-yet-distributed' innovation. Our brain creates the illusion of a unified, coherent sense of self from many interacting neural networks – *from the many, one*. Beyond this, our sense of self is influenced by other people all around us – *from the many, one*; and any useful tool that we wield in our hands is the work of thousands of others – *from the many, one*. Our 'one' depends on others in so many ways.

Species of social insects like ants, honeybees and termites work in a similar way. Individuals from a colony work together closely, each carrying out their specific tasks in a harmonious interplay that enables the coherent functioning of the colony. The individual units cannot function alone. Individuals from worker castes are incapable even of reproducing themselves – they forgo autonomous sexual reproduction and instead work with siblings to ensure the colony is healthy enough to produce reproductive castes once a year. If aliens from outer space were to observe the creatures of planet earth, they would probably see how closely the ants of a colony work together and conclude that an individual unit is actually the single colony. Likewise, if the curious extraterrestrials were to observe the intricate ways in which humans work together and build upon one another's innovations, passing products from many hands and each adding their small contribution, perhaps they would conclude that all humans are in fact one interconnected unit that cannot be accurately analysed as component parts alone. After all, where is the evidence that you could pull a single individual from this tangled tapestry of interactions and they would survive alone? From observation of the deep mutualistic co-dependency of humans this would appear unlikely. And, indeed, when a person is accidentally cut clean away from the lifeline of cooperative interactions they rarely survive. Alone in the outback, the mountains or the tundra, the elements are brutal. After a plane crash, in such wilderness, popular rhetoric is that a good knife is the bare minimum you need to have any chance of surviving; yet in that single object lies concentrated the connections to the whole web of humanity. You can easily imagine Leonard Read's essay being called 'I, Knife'. Take away that object, as well as your clothes and any other items made by others, and you are nothing but a fragile body. Pulled apart from the complex web of human interactions, like an ant plucked from its colony, you would not survive for long. Your existence is contingent on this tangled tapestry of interactions with other people.

9

You thrive when the whole is intact

Right now, the earth is buzzing with human connections. Like a great superorganism, we are linked to each other by intricately directed flows of matter and energy across the Earth's surface, underground and through the airwaves. Over 64 million kilometres of roads are laid across the globe,[1] along which crawl 1.1 billion vehicles carrying goods to where there is demand, just as the blood cells in your arteries carry oxygen to the rest of your body.[2] The world's oceans carry over 560,000 ships, transporting bulk goods and, as you read these words, over 6,000 commercial passenger planes race through the air, suspending 1.8 million people in the sky, zipping between 41,000 airports.[3] And what an awe-inspiring wonder is the nervous system of the earth: millions of miles of electric cables buried under the seabed or strung like fine spider-webs across the landscape. Above all this, 13,000 satellites orbit the planet like a great man-made constellation, transmitting electromagnetic waves in every direction, allowing 7 billion people to chatter away to each other on phone calls, radio, TV and the internet. Reach out your hand into the air and hold it there –

a hundred conversations are flowing through it right now!

All this global interconnectivity needs a highly efficient and intricate network of cooperation between people. It also requires many thousands of different types of resources: metal for cables, buildings and electric components, sand for cement, oil for plastics and food to feed billions of people every day. As we read earlier, our human bodies are open systems requiring regular chemical inputs in the form of food and they convert this into heat and movement energy. With seven billion human engines humming away constantly, that requires a lot of fuel, plus the manufactured infrastructure (buildings, roads, clothes, gadgets) to support these human bodies. As a society, we use over 164 billion kilograms of raw materials every day (that's almost two tonnes per second),[4] and the use of many of these is increasing. The extraction of resources and the waste products their use generates now creates such a strong signature on the Earth's biophysical processes that geologists argue we are entering a new era called the Anthropocene. Although this may pander to the self-important parts of us, proud that humans are a dominant force in the world, it actually marks an era where our relationship with the natural world is increasingly precarious. Many of the resources we use are finite and the waste products from our industries – greenhouse gases, fine particulates in the atmosphere, nuclear waste, microplastics in the oceans, fertilisers and pesticide chemicals leaching from agricultural fields – accumulate in the environment, causing increasing harm to our health and that of other species. Well-regarded medical journals like the *Lancet* and *British Medical Journal* are now making links between human health and large-scale planetary processes, such as climate change, biodiversity loss and air pollution. The threats are often greater to people who are less well-off – those living in poorer countries and in the poorer neighbourhoods of cities.

We are becoming more efficient at using resources – increasingly, we recycle our waste and it becomes the input into new industries (the fabled 'circular economy'), and we are rapidly developing renewable sources of energy like solar, wind and biofuels. But this shift is happening *only to a limited degree*. It doesn't come close to offsetting our increasing demand for resources. Economists talk about how we are achieving 'relative decoupling' between resource use and economic growth – where we have boosted efficiency at using resources, using fewer inputs for every dollar of a country's GDP. However, we are nowhere near achieving 'absolute decoupling' – a point when increasing economic growth does not depend on further resource depletion. In fact, new mines, oil wells and intensively managed croplands are still expanding, with subsequent impacts on ecosystems.

What are these ecosystems that environmentalists are so concerned about? The term was coined by ecologist Arthur Tansley[5] in reference to networks of living organisms connected to each other, and the material world, through flows of matter and energy. At the most basic level, species are connected to each other by 'trophic' interactions, such as predator–prey, herbivore–plant and host–parasite. By the consumption of another organism, matter and energy pass from one body to another. Such flows can be summarised as a biomass pyramid: the biomass of all primary producers, such as plants, make up the bottom layer of the pyramid, herbivores are the middle layer, then predators at the top. Energy and matter flow up the pyramid, as herbivores eat plants and in turn are themselves eaten by predators. In each layer, only 10 per cent of energy is transferred up to the next level, with 90 per cent used to fuel metabolic processes of the organisms, in other words lost from the side of the pyramid as kinetic and heat energy; hence the pyramid becomes smaller the higher up you go.

Besides eating one another, species also interact in more subtle ways. Some influence each other by competing for shared resources, while others work together for mutual benefit. The most obvious examples of species working in harmony are plants and pollinators – insects such as bees benefit from gathering energy-rich nectar from plants and, in return, transport pollen between flowers allowing plants to reproduce sexually. There are many other types of mutualism, from predatory fish restraining themselves from eating up small cleaner fish, allowing them instead to pick parasites from their bodies, to plants that provide cavities in their stems to house ants who, in return, guard them from herbivores. In some cases, mutualisms are so intricate it is hard to see where one species ends and the other begins. Lichens, for example – the crusty organisms growing on the surfaces of trees and rocks – are actually associations between different species of fungi and algae, all sharing the same body. And we have already discussed how in our own bodies, deep inside our cells, we are actually a chimera of ancient free-living bacteria and human cells, plus those casual dependencies we have with the bacteria in our guts.

The outcome of all these interactions between species is

a deep complexity in how ecosystems operate. It means that impacts in one part of an ecosystem can cascade through and affect other parts in unexpected ways. Despite ongoing scientific research, we still lack the capacity to understand most of these relationships, but there are some examples of intensive research demonstrating the amazing potential for knock-on effects in these interaction chains.

In the hills of south-western England, huge numbers of striking Large Blue butterflies were once found, yet over several decades in the mid-twentieth century their numbers started to dwindle, and no one knew why, or how to prevent the decline. In 1979 the butterfly was sadly declared extinct in Britain. When scientists began investigating the intricate interactions that led to its extinction they discovered the species depended on a specific type of red ant living under the soil. The butterflies tricked the ants into thinking their caterpillars were ant larvae so the ants would carry them underground and rear them as their own, allowing the butterfly to emerge just under a year later, well-fed and protected. These red ants were a heat-loving species that thrived in areas where the grass was short and sunlight could warm the soil; but the height of grass in the region had changed, in part due to a decline in the number of grass-eating mammals. Farmers had reduced the number of sheep grazing the fields, while the myxomatosis virus had decimated the rabbit population; without these two species feeding on grass, keeping the height down, vegetation grew taller and denser, shading the soil and making it too cool for the ants. Thus, a very complex chain of events was responsible for the butterfly's fate: the presence of a rabbit virus and grazing-stock decisions by farmers had knock-on effects on vegetation height, which in turn affected ant numbers, which in turn drove the decline of the butterfly. On understanding this, the conservationists were able to implement a reintroduction of the large blue butterfly from Swedish populations, this time ensuring optimal grassland height through careful grazing management. As a consequence, over thirty colonies once again thrive in the south-western English hills.[6]

If slightly changing land management and reducing the numbers of a couple of herbivore species can have such far-reaching effects on an ecosystem, think what reducing thousands of species might do. Rather than pulling at a couple of threads in the tapestry of life and seeing what happens, our crude industrial activities – intensive farming, trawler fishing, mining and bulk transport – are essentially tearing away great swathes of the tapestry, with unknown consequences. We have started to undermine the foundations of the pyramid of biomass mentioned earlier. The energy and materials produced by plants and other organisms (those 'autotrophic' organisms capable of generating energy for themselves from sunlight or inorganic chemicals in their surroundings) are usually available for other species, such as herbivores, who then support a varied range of predators and parasites. Though we humans are just one species of an estimated 8.7 million on the planet,[7] we appropriate for ourselves over *one quarter* of all the materials produced by plants across the whole earth.[8] Add to that the harvesting of higher tropic levels through overfishing and hunting, plus the additional impacts on species through pesticides, chemical fertilisers, microplastics, acid rain and many other environmental pollutants, and the scale and impacts of our human enterprise are enormous.

We have allowed this damage to the Earth's ecosystems because we have often seen nature as something external to us, something to be conquered and used. It is becoming increasingly clear, however, that continued destruction of nature will simply accelerate our own destruction: the more environmental damage we cause, the more susceptible we become to extreme weather events, invasive species, pandemic disease and food insecurity. Seeing these problems on the horizon, environmentalists in the 1990s decided to change tack in their strategy to conserve the natural world. Previously, environmental causes were championed by urging us to protect iconic species, such as the panda and polar bear. Raising the plight of these beautiful creatures was hoped to be enough to secure their survival; yet it was not – calls to protect nature for its own beauty were insufficient to stem the tide of destruction from a blind and faceless global economic machine. As a consequence, many in the environmental movement switched tactics to highlight the *instrumental* value of species, measuring the many useful things they do for us – pollinating crops, controlling pests, protecting soils from erosion, regulating the climate and purifying water, to name but a few benefits. Conservationists called these ecosystem services, explaining that, in the absence of wild species, using human labour or technological innovations to replace them would be hugely expensive or impossible. Nature provides these services for free, but their true value is not accounted for in our economic system – when human activities cause damage, such as deforestation preventing the ability of an ecosystem to purify water, the costs are ignored. At the national level, measures of a country's wealth such as GDP (Gross Domestic Product) can go up, leading governments to declare success, while the value of nature is being destroyed.

The solution, as conceived by a set of new environmentalists, was to better capture these values of nature in decision making – in the economic jargon, to 'internalise the costs' of nature's destruction. Much of the language adopted by this new conservation agenda was borrowed from economics. Nature was referred to as 'natural capital' comparable to the other forms of human and financial capital that economists were already familiar with. Broadly, the idea was that if you can't beat them, join them – capturing nature's value in real monetary terms might just convince decision-makers that its protection simply made good economic sense. Conservation was no longer about iconic species and the beauty of the natural world; the new fashion was for market economics and high finance.

Yet, this could turn out to be a big mistake. If the root cause behind our destruction of nature is a failure to recognise it as essentially part of us, then an approach which treats it as an abstract commodity is problematic, because it divorces us from nature further still. The new paradigm of natural capital assumes we don't need to correct our self-identity, or the way we behave as consumers – we can simply correct the economic system so that any costs to nature are internalised. Yet, without dealing with the root cause of our prolific consumption, the impacts on nature will simply ratchet ever upwards.

Although many major global organisations such as IUCN – International Union for Conservation of Nature, the world's largest global environmental organisation – are now beholden to the concept of natural capital, not everyone is sold on the idea. Many people, including those from indigenous cultures with strong traditions that recognise our dependency on nature, reject such commodification of the natural world. They prefer a holistic approach. As Professor Sebsebe Demissew of Addis Ababa University in Ethiopia explains: 'In such cultures, it makes no sense to place a monetary value on a forest or a river because they are part of the whole body. It's like saying to a human: "What price, your limb? Or what price, your kidney?"'9

This debate has been reflected in the undertaking of the world's biggest ever assessment of nature. The Intergovernmental Science-Policy Platform on Biodiversity and Ecosystem Services (IPBES) is a huge global cooperative effort to understand the status of our plants and animals, akin to what the Nobel-award winning IPPC (Intergovernmental Panel for Climate Change) has done for climate change science. The process of bringing together hundreds of scientists and policy makers from very different cultures and understandings of our human relationship with nature has not been an easy ride.10 Many participants rejected an initial framing which used strong economic language – nature providing 'ecosystem services' to humans – and, eventually, a more inclusive approach was adopted. The assessment now recognises the central role that culture plays in defining links between people and nature. As several of the IPBES team authors explain, the new approach 'explicitly recognises that a range of views exist. At one extreme, humans and nature are viewed as distinct; at the other, humans and nonhuman entities are interwoven in deep relationships of kinship and reciprocal obligations'.[11] This reflects a promising change in direction from the dominant economic approach to conservation which has pervaded over recent decades. It offers a greater likelihood of resolving the apparent disconnect between nature and self that leads to 'selfish' decisions where we benefit ourselves in the short-term but at the expense of the natural world which sustains us.

A closer relationship with nature, which recognises our intricate interdependency upon it and integrates our self-identity into it, is essential for us to thrive on this finite planet.

It is thought that 90 per cent of species (over 7.5 million) on earth still await description, despite over 250 years of taxonomic categorisation.[12] It is idealistic to think that, through reductionist economic analysis, we can understand what each species does and then incorporate its monetary value into our economic markets. Instead, it is essential to take a precautionary approach to minimise our detrimental impacts on the natural world. The American author and environmentalist Aldo Leopold once said that 'to keep every cog and wheel is the first precaution of intelligent tinkering'. Yet, the IPBES global assessment shows that, as of 2019, three-quarters of our global environment on land and about two-thirds of the earth's marine environment have been significantly altered by human actions. Plastic pollution has increased tenfold since 1980 and other pollution has produced more than 400 'dead zones' in the world's oceans, totalling over 245,000 km^2 – a combined area greater than that of the United Kingdom where conditions are so toxic that marine life is almost totally absent. In total, over one million species are threatened with extinction globally. This extensive destruction of biodiversity is far less than intelligent tinkering. We are un-weaving the tapestry of life, where every thread is deeply connected to another, and where the loss of one species can cause cascade extinctions in others – like the 'butterfly effect' that Edward Lorenz found in his interconnected weather patterns yet, here, linking real butterflies along with millions of other species. Through these extinction cascades, we risk losing precious, yet undiscovered, plants and creatures that could hold the next medicinal cures or control new pests. Ultimately, we risk the extinction of our own species.

An ecosystem is a complex and awe-inspiring dance of interactions, and we humans are deeply interconnected as part of these networks. As we have learned, we do not have distinct bodies, we are an open interface between inner and outer ecosystems – between the non-human cells within us, and the wild species surrounding us – both of which are essential to supporting our existence. We are not isolated minds operating atomistically, we are part of a remarkable cooperative human initiative that has created technologies and a global economy of dizzying complexity. But we now know that this human industry threatens the very basis of our life support systems. More than ever, we need to broaden our perspective of self-identity to prevent the long-term self-harm caused by our actions. The naturalist Charles Darwin described the complex ecological relationships between all species as a 'tangled bank'.[13] When we destroy these species and degrade the natural world, we are stealing from ourselves. The ultimate cost of plundering nature's tangled bank may be our own species' survival.

PART THREE

OUR SELF DELUSION

See how beneathe the moonbeams smile
Yon little billow heaves its breast
And foams and sparkles for a while
And murmuring then subsides to rest,
Thus man, the sport of bliss and care,
Rises on times' eventful sea,
And having spent a moment there,
Thus melts into eternity!

Thomas Moore / Hannah Dracup[1]

When we try to pick out anything by itself we find it hitched
to everything in the universe

John Muir, *My First Summer in the Sierra*

10

The Big White Lie: the trick behind our sense of self

Have you ever told a little white lie to someone? Perhaps you made up an excuse to avoid a social commitment because you were too worn out, but you didn't want the host to think you just couldn't be bothered, or maybe you omitted to tell someone the truth to avoid hurting their feelings? Although we are taught as children to always tell the truth, we sometimes tell small lies when we perceive there to be no harm done or we feel we may be doing some good. We ultimately tell little white lies for the greater good (or so we tell ourselves), and they permeate the fabric of our social culture. However, like many aspects of human invention, Nature got there first. Evolution has set up one of the biggest white lies of all time, and we humans are truly hoodwinked.

We are susceptible to certain illusions. Because the sun always rises in the east and sets in the west, it appears that the Earth stays in one place and the sun moves around it. This seemingly sensible inference held sway in human understanding for thousands of years. It was not until the sixteenth century when Renaissance mathematicians and astronomers such as

Nicolaus Copernicus, Johannes Kepler and Galileo Galilei finally managed to overturn the status quo and the heliocentric model (where the sun is at the centre of the solar system and the planets revolve around it) began to be widely accepted. In a similar vein, humans previously thought of themselves as being created by God in 'his' image before biologists such as Darwin and Wallace in the nineteenth century finally managed to convince many people that we are, in fact, descended from ape-like ancestors. In both cases, human beings were knocked from their pedestals, where they had been raised up as unique, created differently to everything else and central to the universal order – all of which were idealistic fantasies. In reality, Earth circles the sun, just like every other planet in the solar system, and in the words of the musician Brian Eno, we humans are not at the top of the pyramid of the web of life but 'just another species, in the innumerable panoply of species, inseparably woven into the whole fabric (and not an indispensable part of it either)'.[1]

Such home truths can be uncomfortable. Even now, in the twenty-first century, a significant proportion of people deny the fact of our evolutionary history.[2] Although on both accounts (heliocentrism and evolution) the public majority is now just about in line with the scientific facts. There is one last pedestal, however, that we humans are stubbornly clinging to – the apparently sovereign independent identity at the centre of our personal universe.

Yet this is a just lie that we tell ourselves. We might like to believe that at the core of our being is a kernel that is our inner self. Independent from the rest of the world, it constitutes our own private identity, the essence of who we are and what makes you uniquely 'you'. Reasonable as this may sound, it is in fact a clever, almost seamless, deception. The truth is probably something like this: our brains evolved to support minds that falsely believe they are independent entities and while our modern culture often celebrates and exaggerates this individualistic perspective, it is, quite simply, a cruel illusion.

Can you truly remember what it felt like to be a one-year-old child? You cannot because that person was not really 'you', at least not the same person you identify with right now. Inside the mind of that child there was no constructed self to make sense of the events that happened to them. Psychological studies of babies and very young children show they do not have a 'Theory of Mind', which would allow them to recognise that other people have their own minds too. Ingenious experiments have been designed to show this. For example, one experiment involves a young child and two dolls (Sally and Anne), along with a marble, a basket and a box. The experimenter shows one of the dolls, Sally, playing with a marble which she puts in her basket before leaving the room. The experimenter then takes the other doll, Anne, and shows her moving the marble from the basket to the box. When Sally the doll is brought back into the room, the child is asked where Sally will look for the marble. Children under the age of about four generally point to the box where they know the marble to be. In contrast, older children point to the basket, understanding that Sally will hold a different, albeit false, belief as to the marble's location. The conclusion is that very young children are unable to understand that the minds of others are different – that other people can hold independent beliefs from their own. Instead a discrete sense of individual identity only develops as the toddlers grow up.

With the genesis of an identity comes benefits: having a Theory of Mind is useful in that it allows us to predict the beliefs and intentions of others, be they benign or malevolent. For prehistoric humans, it would be a great help to know whether the individual approaching you with their hands behind their back was holding a club to attack you or a gift of friendship and alliance. Maintaining an apparently coherent identity also allows us to tie together memories, helping us to perform better based on past experiences. For example, if an ancestral human being could remember a new technique they had discovered to scale a tree and obtain fruit, then a coherent memory, recalling the action and its benefit, would allow them to repeat the new trick and show it to others. Our memories and our identities are essential survival tools, and evolution has sharpened the subjective sense of independent autonomy in line with the success it has brought us. However, just because we have this *sense* of identity, this feeling that we are the individual keeper of a great storehouse of personal memories and experiences, does not in any way mean it is a genuine objective entity.

Take our memories: things that happened to 'us' which we have encoded in our minds and that make us who we are. As adults, we appear to be able to integrate these memories and beliefs into a coherent sense of self-identity, but it is really not as coherent as we think – look closer and you will see the cracks appearing. As early as the 1930s, researchers began to realise memories are not exact copies of past events but are reconstructed like stories. Each time we recall these stories, they are liable to be changed. In asking people to recall a video of a car accident, for example, if researchers ask leading questions (such as, 'did the white car jump the red light?' when in fact there was no white car), in many cases people start to believe these statements and integrate them into their memories when recalling them later. Similarly, if people are shown faked photographs of themselves as children in hot-air balloons, many will go on to recall the event in detail![3] It can be quite a shock to people when they are made aware of the truth. Think of a memory from your childhood which is important to you. Whether you recall it like a short video or as a photograph-like snapshot of a moment, isn't it jarring to think that many of the details may not be true factual representations, but rather our brains filling in the missing details. As Bruce Hood, a developmental psychologist, explains:

> Our self-illusion is so interwoven with personal memories that when we recall an event, we believe we are retrieving a reliable episode from our history like opening a photograph album and examining a snapshot in time. If we then discover that the episode never really happened, then our whole self is called into question. But that's only because we are so committed to the illusion that our self is a reliable story in the first place.

If our memories are malleable is this way, and our identities are bound up in our memories, then it follows that our identities are fallible too. And we not only remember poorly, we often embellish the details, especially when they pertain to our own abilities. For example, ask people how good they are at driving, or how good their sense of humour is, or any other attribute that is important to their sense of self-worth, and over 50 per cent will say they are better than average. The maths just don't add up. They demonstrate that people suffer from an inherent cognitive bias regarding their superiority to others.[4] A good part of what we believe ourselves to be is simply not true.

The lies we tell ourselves are context dependent too – our sense of identity depends on where we are and, crucially, who we are with. Two well-known philosophers, William James and Friedrich Nietzsche, born two years apart in the 1840s in America and Germany respectively, both recognised this. Nietzsche described how our selves are essentially reflections of how others see us, while James described how we have as many selves as the people we interact with in different social situations. Some decades later, in 1902, Charles Horton Cooley coined the term the 'looking glass self' to express the way the self is shaped by the reflected opinions of others around us. Cooley was a sociologist who often took an empirical observational approach, for example, developing insights from observing his own children.[5] He noticed how we build and continually develop our self-image based on other people's views (or at least how we perceive their views of us to be). This led him to conclude that society and the individual are not separable phenomena but different aspects of the same thing.[6]

Our identity is not only contingent on connections with the

outside world, it is also dynamic as this social context changes. Neuroscientist Susan Greenfield suggests our identity is best seen as an activity rather than a state. Based on research into clusters of neurons in the brain (neuronal assemblies) which are rapidly activated in changing patterns to generate consciousness, Greenfield concludes that identity is not some kind of solid object or property locked away in our heads. Instead, it is a type of subjective brain state, a feeling that can change from one moment to the next.[7] This shifting of self-identity is implicitly acknowledged in some languages, such as Japanese, which has different words for 'I' depending on who we are with. In the religion of Buddhism, through deep introspection and meditation, scholars maintain there is no unchanging, permanent soul within us, leading to the so-called principle of *anatta* or 'non-self'. These conclusions align closely with the 'bundle theory' advocated by Scottish philosopher David Hume, where there is no underlying object which is our true self but only properties that are constantly changing and context dependent. These changing contexts modify the story we tell about ourselves, meaning that the self that we 'predict into existence' as the neuroscientist Anil Seth puts it, is constantly in flux. Daniel Dennett, the contemporary American philosopher and cognitive scientist, expresses this eloquently, saying that we have no self at our core, but instead a self 'emerges as the centre of narrative gravity'.[8]

If, for hundreds of years, philosophy has pointed to the fact that we have no independent unchanging self, and in more recent decades the sciences of psychology, neuroscience and cognitive science have all come to the same conclusion, why, then, do we maintain such an unshakeable subjective feeling that we do? Of course, to know a truth theoretically is different to fully experiencing it and acknowledging it. Even if we are occasionally admitted a brief insight into our true nature. As much as we try to step outside of our minds to see this objective truth, we are rapidly pulled back into our mental cockpit, where we subjectively experience the world as a seemingly coherent 'me'.

This is because evolution has expressly crafted the design of our bodies and our minds to see the world from this perspective, and it has been a spectacularly successful adaptation. The illusion of self allows us to navigate our bodily vehicles in a way that they are less likely to be eaten by predators, more likely to find food and shelter, and more likely to achieve higher social status, mate and produce the next generation of self-deluded but very successful humans. The cognitive neuroscientist Matthew Lieberman emphasises the importance of this last factor, sociality. He suggests the evolution of the self is a 'sneaky evolutionary ploy' to ensure the success of group living. This social hypothesis is based on observations that during much of the 'spare' time of the brain, when it is not carrying out autonomous cognitive tasks, it is not resting but rather very active. When we are daydreaming or sleeping the brain is actually intensely active, and it turns out the activity is in the same areas of the brain that control social interactions.[9] On a different tack, the German theoretical philosopher Thomas Metzinger suggests a fictional self is necessary for important functions like reward prediction, highlighting how it only makes sense to plan for future success when you have a strong feeling that it's going to be the same entity that gets the reward in the future. He speculates that the intense activity of the brain during daydreaming and sleep is to provide 'autobiographical self model maintenance', meaning the brain works hard to maintain the self-deception that we have a persistent personal identity across time.[10]

Whatever the reason, it is clear that evolution has designed a self-deception that has well and truly hoodwinked us. Our genes create a mind which is amazingly successful in maintaining the illusion of an independent, autonomous person. Evolutionary biologist Mark Pagel phrases it strongly:

> If the last sixty years of experimental psychology, personality, evolutionary, and neurological studies have shown us anything, it is that the minds that have proven useful in that struggle are far more bewildering than might be expected. For one, the inner 'I' that you think you know so well probably doesn't exist. It is an illusion, the construction of a mind, which is itself the construction of its genes, genes that have been selected to produce brains that further their ends. Those brains will use false beliefs, copying, lies, deception, self-deception, and just about anything they can lay their neuronal hands on to promote our – and consequently their – survival and reproduction.[11]

At the start of this chapter, I described how we humans, whose clever cultural graces include those well-meaning untruths – little white lies – have been completely outdone by evolution, which has pulled the wool over our eyes so well that we cannot see the interdependence of our selves with the outside world. Unlike a little white lie, however, this is quite a hefty one; although we can now see it still warrants being granted the merits of being 'white' on the basis that we, the deceived party, benefit from the deception. By definition, natural selection crafts new designs that provide benefits to their bearers. The feeling of a coherent, unchanging inner self is an illusion, yet it has been an *adaptive* illusion – a survival tool allowing humankind to thrive by developing sophisticated behaviours and productive social networks, where we work together for the benefit of all. Although Hume's ever-changing 'bundle self' may be the ultimate reality, can you imagine trying to get on in life without a coherent self-identity? In today's fast-paced world, with its complex culture, the need is greater than ever. We interact with so many people nowadays, and just think about trying to feed yourself without the concept of a self to interpret your hunger and satisfy it by tying together memories of where and how to get food. Navigating the complexity of what to buy, we constantly query our self-identity: 'Am I allergic to any ingredients in this one? Can I afford this one?', and so on. Without some semblance of coherency in our self-reference, our lives would be a mess.

If the self is an adaptive survival tool, we might think that perhaps we should be careful in trying to lift the veil that evolution has placed over our eyes. Would entirely abandoning our self-identity lead us to become an amorphous and ruinous wreck? Sufferers of the rare mental illness known as Cotard's syndrome deny they lack an independent self-existence and certainly struggle to get on well in the world. The syndrome is accompanied by despair, depression, self-loathing, and sometimes intense delusions making sufferers think they are 'walking dead'. There are other dramatic ways to lift the veil of illusion. Recent research shows how the drug LSD inhibits the experience of self by increasing connectivity across our brain networks. The drug seems to disrupt the synchronised brain networks that fire together while the brain is at rest in 'default mode' and maintain an autobiographical sense of self agency. Reduced activity in these brain networks under LSD, and other psychedelic drugs like psilocybin, correlates with volunteers reporting a disintegration of their sense of self, or ego – an effect known as ego dissolution. Instead of running these networks, LSD seems to instigate broader patterns of global connectivity across the brain, which could explain the reported enhancement in creativity under the drug, from linking up neural networks that wouldn't normally be connected.[12] With excessive dosages the loss of self-control can be dangerous (though tales of people who took too much LSD and jumped from the top of buildings thinking they could fly might just be urban legend). A drug-free alternative is meditation, and there is a current trend in self-help culture, which is rapidly becoming mainstream, for mindfulness meditation. Clear as the long-term benefits of such introspective approaches are (and there do appear to be many), there are an increasing number of reported cases where, in certain people, it can lead to unexpected extreme anxiety. According to two separate surveys, conducted between 2013 and 2017, over a quarter of people who regularly meditate have experienced unpleasant feelings as part of the practice including disturbing feelings of fear, dread, or terror.[13]

Perhaps we need to be more selective in deciding when and how to peel away the illusion of our selfhood. Having a self is essential for us to get on in life – it is the reason such a perspective evolved in the first place, therefore we certainly don't want to (and couldn't easily anyhow) completely discard this useful perspective. It is a survival tool that helps us navigate life successfully, as essential as a penknife if we were lost in the wild. However, knowing our sense of self to be an adaptive illusion, there may be occasions when we wish to use our objectivity to pierce this falsehood, when circumstances benefit us doing so. When the 'I' in our heads gets carried away in narcissistic monologues, it may be beneficial to curb the garrulous tendencies of this actor upon life's stage and shift our perspective to see the dependency on the rest of the cast. After all, it is everyone and everything around us that makes us what we are.

11

Your seat has a restricted view

A conversation begins with a lie. And each
speaker of the so-called common language feels
the ice-floe split, the drift apart
as if powerless, as if up against
a force of nature

Adrienne Rich, *Cartographies of Silence*

You are nature. You are a hominid ape. You are in the
world and the world is in you. Everything connects!

Matt Haig, *Reasons to Stay Alive*

What do the two quotes above have in common? They seem
diametrically opposed to each other, one highlighting our
intimate connectedness to the world and the other our perennial isolation. In common, however, they are both extreme
views compared with our normal perspective. For most of us,
most of the time, our point of view lies somewhere in between
these two extremes of interconnectedness and isolation. Of
course, we are all different, and people occupy different points
along this continuum, driven by a combination of our genes

and, of crucial importance, the cultures we are brought up in.

There are differences between human cultures in their tendency to see objects and events in isolation versus recognising the interconnections between them. In particular, there are marked differences between Western cultures (such as Europe, USA and British Commonwealth countries) versus East Asian cultures (such as China, Korea, Japan).[1] Psychologists have found that, on balance, people from Western cultures are more likely to see events and objects in isolation, including themselves, which leads to 'individualistic' attitudes. In contrast, people from East Asian cultures are more likely to perceive things and events as an inextricable part of a broader context and, with regards to self-identity, they tend to adopt 'collectivist' attitudes. This is a very broad stereotyping. The quotes at the start of this chapter highlight how people with a Western background (such as the author Matt Haig) can still adopt attitudes focusing strongly on interconnectedness. Incidentally, Matt Haig's book *Reasons to Stay Alive*[2] describes how his life has oscillated between intense depression, characterised by feelings of deep isolation, versus intense joy at the interconnectedness of the world. So, stereotyping Western and East Asian mindsets as if everyone in those groups are the same is a gross simplification. However, that does not preclude there being broad *average* differences between groups. In fact, differences in human perception between the two biogeographic-cultural origins are supported across several different lines of evidence, some of which I will describe, although as cultures mix the broad differences between these poles are becoming increasingly blurred over time.

Meet James, a stereotypical Westerner. As someone from a historical culture of individualism he is more likely to group objects into discrete categories, he views other people as having attributes independent of their circumstances, and he learns words that are nouns more easily than verbs. Overall, with respect to identity, James has a strong sense of personal agency and autonomy.

Now, in contrast, meet Jiang; a stereotypical East Asian, Jiang will tend to focus upon relationships between objects and events, he will consider the context in which objects occur, and learn verbs faster than nouns. With respect to his sense of identity, Jiang places less value on his individuality and higher value on the social group to which he belongs.

The psychologist Richard Nisbett in his book *Geography of Thought* speculates upon how these cultural differences in systems of thought stem from many centuries ago. Western culture owes much to the Ancient Greeks, who placed great value upon individual freedom and the capacity for individuals to express themselves. They openly encouraged competition across many domains, for example through Olympic sport, gladiatorial combat or dialectic discourse (the latter emphasising the value of logically reasoned arguments to settle different points of view). Literature, from *Odyssey* to the *Iliad*, features gods and heroes with strong individual personalities, and the Ancient Greeks celebrated and revered a sense of personal agency free from constraints.

In contrast, early Chinese cultures aspired towards social harmony, with individual identity strongly embedded in social roles. For example, Confucius (551–449 BC), one of the most influential figures in the history of Chinese thought, endorsed the concept of *ren* – a morality cultivated through education in social norms such as courtesy, ceremonies and good manners. According to Confucius, moral education should start with *Xiao* (filial piety) and *di* (brotherly love and respect) and extend to *Zhong* (loyalty to one's lords and superiors). In this way, both family life and the State attain peaceful order. The moral code of Confucius prescribed the norms of human relations and, beyond these, the concept of individuality had little meaning. As the philosopher Henry Rosemont summarised: 'For the early Confucians, there can be no me in isolation, to be considered abstractly: I am the totality of the roles I live in relation to specific others ... Taken collectively, they weave, for each of us, a unique pattern of personal identity, such that if some of my roles change, the others will of necessity change also, literally making me a different person.' This focus on social harmony, group identity and neglect of the individual has remained a key strand in East Asian cultures for many centuries.[3]

The broad difference in self-identity and the sense of personal autonomy between cultures outlined here is apparent not only from the analysis of ancient texts. Amazingly, even in the present day, there are measurable differences between people from different cultures, which have been explored in elegant behavioural psychology and cognitive science studies. There are several tests, for example, for a phenomenon that is referred to as 'field dependency', in which the perception of an object is influenced by the environment in which it appears. Imagine being asked to identify a simple geometric shape embedded in a complex background pattern. It turns out that on average people from Western cultures are better at finding the shape, as, incidentally, are autistic people.[4]

Another test is designed to investigate whether participants take notice primarily of the substance or the shape of objects. American children and adults when shown an object, such as a sphere made of cork, and asked to pick other objects that are similar tend to pick ones that have the same shape regardless of the substance (for example, a rubber cork). In contrast, Japanese individuals pick objects of the same substance, even if the shape is different. In a different test, participants are shown animated images of fish tanks. When asked to recall what they saw, Japanese participants tend to pick out details of the background and describe the whole environment, compared with Americans who focus on individual fish. So, in general, it appears people from Western backgrounds more easily focus on distinct objects, while those from East Asian backgrounds are much better at considering objects as part of a broader context in which they occur.

Cultural differences in how objects in the environment are seen in isolation or as part of an interconnected whole are also mirrored in experiments designed to investigate how human beings regard themselves. When asked to describe a recent event, American children refer to themselves three times more frequently than Chinese children.[5] When asked to describe themselves, the Americans frame responses in terms of absolute personality traits or roles (for example, 'I am an excitable person') whereas Japanese participants tend to identify the context that a personality or behaviour is expressed in ('these type of situations make me excitable'), and they refer to other people twice as often in their personal descriptions.[6]

Do all these differences really matter? Well, if we care about things like personal happiness and humanity's ability to navigate successfully through the Anthropocene, then the answer is absolutely yes – our perceptions fundamentally influence how we interpret the world, ultimately determining both our individual behaviour and the structure of our institutions. In terms of the two poles of seeing objects in isolation (typified in modern Western cultures) versus perceiving their interdependency (typified in modern Eastern cultures), one can argue that encouragement of the former approach has been at least partly responsible for fuelling rapid technological and scientific advances of the modern world. The ability to first see objects in isolation and then categorise them into common groups (requiring de-contextualisation and abstraction) in what might crudely be called a reductionist approach, has been crucial to advances in philosophy, science and technology. It has allowed us to make ordered sense of a world that William James once described as a 'blooming, buzzing confusion'.[7] By categorising the jumble of objects in the world, we can develop scientific rules about the attributes of materials and how they behave in different situations, allowing us to develop new technologies and arts. Such advances have also been catalysed by our individualist culture, in which competition between individuals is celebrated and objective discourse is welcomed. This discourse ranges from informal gathering of individuals in dialectic debate, to the development of critical-thinking institutions such as universities and academic journals, which encapsulate these values. In contrast to this spirit of individual agency and competitiveness, innovation may be hindered in strict hierarchical cultures, where questioning of superiors is strongly frowned upon and being humble rather than aspiring to stand out from the crowd is seen as correct behaviour. Because of the clear potential of a reductionist and individualistic worldview to drive scientific and technological innovation, which then drives economic development, Western culture has spread across much of the globe in recent decades. This is certainly not to disregard the important scientific advances made in Eastern countries. Papermaking, printing, the invention of gunpowder and the compass, to give a few examples, were all invented in the Far East well before modern Western culture had developed. The main point is simply that the celebration of rational thinking and the reductionist scientific approach (for example, as typified during the Enlightenment period of the eighteenth century in Europe) contributed greatly to the rapid acceleration of scientific progress and the fast pace of technological development we see today. Common wisdom, however, suggests that anything taken to extremes can become bad. The reductionist-individualist worldview is now such a dominant paradigm, with the majority of countries increasing in individualistic values over recent decades,[8] and numerous problems are emerging. Many of these can only be solved by a rebalancing of perspective to focus on the interdependency of objects and events.

What are these so-called problems that I argue should prompt a fundamental restructuring of our historically successful worldview? There are several: personal happiness, disengagement from nature, and an inability to solve the big sustainability challenges of the twenty-first century. To deal with the first, research suggests loneliness is an increasing phenomenon, despite our increasing digital interconnectedness through social media (or perhaps partly because of this). The modern worldview, which celebrates and encourages individuality, downplays our intimate dependency and attachment to others. Yet, at our core, we are social animals – close relationships bring us emotional rewards and satisfaction in life. Social connectedness and support are associated with lower levels of stress hormones and stronger immune function,[9] while being excluded from a social group activates the same areas of the brain associated with physical pain.[10] Modern technologies, such as social media, offer only a mirage of social support, because although on paper (or rather *in silico*) we may have more friends, these provide far fewer of the valuable health and well-being benefits derived from face-to-face contact in close social relationships.[11] The problem of losing touch in our personal connection with others and the world around us is exacerbated in the modern world, where we are encouraged or forced to move jobs to distant places (to pursue individual career success) even if this isolates us from close friends and family.

A second emerging problem with the reductionist-individualist worldview is a phenomenon that has been called 'nature deficit disorder'. This refers to a process of disengagement with nature, especially among children, which psychologists fear could lead to behavioural and physical health problems in later life. Some suggest this is a quite recent phenomenon caused by a combination of factors that limit children's time spent playing outside: increasing fear of strangers, the emergence of screen technologies (TVs, iPads, smartphones), reduced green space in urban areas. However, others suggest the disassociation between humans and nature has actually been growing gradually over many generations, driven by our tendency to see humans and nature as separate things.[12] The more we see ourselves as isolated individuals, the less we enjoy ourselves as part of a greater nature. Whatever the origins of the problem, the physical and mental health benefits of engaging with nature, and the costs of a lack of such engagement, are becoming increasingly well documented. They include lower achievement at school, poorer mental and physical health and under-developed social skills.[13] The UK nature conservation charity, RSPB, reported that only 10 per cent of children in the UK played regularly in natural places in 2009, compared to 40 per cent in the 1970s. In a 2013 survey, they assessed levels of nature connection as insufficient for three-quarters of eight- to twelve-year-olds.[14]

Finally, a third problem with the reductionist-individualist worldview is that, even though it has allowed great scientific advances, there are limits with what can be achieved using such an approach. The big problems of our time, things like climate change, pandemics, social injustice, poverty, natural resource depletion, biodiversity loss, are all systemic problems that cannot be solved solely with reductionist logic. In many cases the problems are linked to each other and require a more holistic approach. For example, biodiversity loss is driven by our need to grow sufficient food and also impacted by climate change, which in turn influences ocean acidification and drives mass human displacement, which then impacts social attitudes towards migration, which in turn influences geopolitical interactions, world trade and so on – the issues are deeply interconnected and technological fixes alone are insufficient to address them. They are sometimes called 'wicked' problems, because they involve complex dynamic relationships between different actors, often across large spatial and temporal scales. To deal with them, our traditional reductionist approach reaches its limits and a new perspective is needed that comprises a *holistic* analysis of our highly interconnected environmental and social systems.

Holism is defined by social scientist Nicholas Christakis as the recognition that 'wholes have properties that are not present in the parts and not reducible to the study of the parts'.[15] For example, to understand the subjective experience of memory, we cannot simply study the structure of neurons. Similarly, to understand ecosystems dynamics we need to know more than just the type of species present. For the last few centuries, what Christakis calls the 'Cartesian project' in science has broken matter down into ever smaller parts in the pursuit of understanding (cell biology, molecular biology, quantum physics). Yet, putting the pieces back together to understand the whole system is often very difficult. In a critique of reductionism, the German biochemist Frederic Vester explained how scientists often study isolated parts of a system in absolute detail, but then completely ignore interactions with other parts of the system.[16] Vester suggested that understanding all the parts of the system, albeit crudely, and, crucially, recognising how these parts link up will give more accurate results in predicting system behaviour.

Encouragingly, this holistic approach is starting to emerge in scientific fields such as systems biology, complexity science and the study of networks. Yet, looking back over history, many philosophers, especially those from ancient Eastern traditions centuries ago, already emphasised the importance of holistic approaches. Some ancient Chinese scholars believed objects are so altered by their context that general rules were not helpful for understanding and manipulating them. This refusal to see objects in abstract isolation may have slowed technological development compared to a reductionist approach, but it now seems highly relevant. Unless we can learn how to view and analyse the world in this way, we will be helpless in the face of huge challenges like climate change and resource depletion, and that ultimately could mean starvation, war and misery for billions of people.

One might question the extent to which the problems outlined above are directly due to a reductionist-individualist worldview, as opposed to simply challenges related to modern living in general. This is a fair question, but there are many ways that the modern world could have unfolded (and still can unfold), and this is contingent upon our personal and aggregate collective worldviews. In subsequent chapters, we will explore how our personal worldviews, in particular reductionist-individualist perspectives, are fundamental in shaping our modern world. Although it is not the only explanatory factor for the wicked problems of our time, it is one of the fundamental root causes. As an example, a proximate cause of biodiversity loss in the tropics is habitat destruction for palm oil plantations, but the ultimate drivers are consumers relentlessly purchasing products containing unsustainably produced palm oil, because they fail to recognise the far-reaching consequences of their actions that extend across the globe. Later in this book, we will touch on the above problems further and investigate potential solutions that arise from moving towards a different worldview.

Modern humans are sometimes in danger of being arrogant in thinking we have carried through all useful knowledge from the past, and now simply add further to it with each new cultural innovation. It is worth considering that some past cultures had a very different worldview, which emphasised the intimate interconnectedness of the world to a much greater (and more accurate) degree. A central tenet of Buddhism, for example, is *Pratītyasamutpāda*, which posits that nothing exists in and of itself but is rather dependent on multiple causal factors. It is clear then that systems thinking has been around for a long time. However, we persist dogmatically with our reductionist-individualist worldview, even in the face of huge systemic problems. It is as if we are still asking 'how does the world fit together?', studying it piece by piece and trying to prise it apart to solve our problems, when what we really need to do is step back and acknowledge the seamless nature of the interconnected world – it is not formed from a jumble of isolated parts like some giant mechanical machine, but is a dynamic system of deeply interdependent components. The two viewpoints (reductionism and holism) are like an optical illusion, similar to the 'Necker cube' optical illusion, based on a simple line drawing of a cube that can be perceived as either 'going into' or 'coming out' of the page.[17] Our perception often flips between these two ways of seeing the cube, but it is impossible to hold the two interpretations simultaneously. The mathematical biologist Marten Scheffer describes how 'snapping' between alternative interpretations happens on multiple levels, not just the interpretation of images, but also between complex theories and worldviews. Perhaps, as a human society, we have fixated on one way of looking at the world for too long (studying limited parts, including ourselves, in isolation), and we are now stuck in this mode of perception. It constrains our ability to see the world from the opposite viewpoint, which recognises the world as a series of dynamic inter-relationships. Unfortunately, unless we can shift our perspective, it is unlikely we will persist for long as a successful species on this planet. We need to explore new ways of thinking about the world and our place in it.

12

Your sense of individuality could be dangerous

So far we have considered how our self – our ego, the little me in our heads – is an illusion. The idea that we are independent, discrete entities is a falsehood created by evolution and exaggerated by our modern culture. We have explored how, over the last several thousand years, and especially in Western cultures, humans developed an increasing tendency to perceive the world in terms of distinct, abstract categories, while also developing a more acute sense of individuality. Of course, both these phenomena are related – the self is the ultimate abstraction, the primal categorisation of objects into 'me' versus 'other'. Such a distinction, by definition, raises a boundary between you and the world that, unless resolved, leaves us trapped in the illusion of isolation.

It is interesting to consider how our sense of individuality has evolved over time, because the surprising thing is, humans have known the truth of our interconnectedness well before neuroscientists and psychologists revealed it though science. With penetrating insight, the Buddha (at least 2,400 years ago)[1] is recorded as stating: 'Our greatest illusion is that we ourselves are an enduring ego, which passes through our life-time ... there is no enduring "I". "I" is just a convenient label for a series of interconnected events.'[2] As Yuval Noah Harari points out, medieval noblemen did not believe in individu-alism, rather someone's worth was determined by what other people said about them.[3] People would go to great lengths to protect their reputation and family honour, to the extent of entering duels and risking their own lives. Similarly, until recent times, the debts of individuals were often taken on by other family members in order to clear the family name. Main-taining the good repute of a family was an absolute priority. And we have seen several examples of how in many Eastern cultures people identify themselves through their position in social hierarchies, in relation to other people around them. So humans have previously been collectivist in their perspective, yet this seems subsequently to have been lost in many modern cultures.

In a complete turnaround in perspective in recent times, psychologists like Sigmund Freud have promoted the idea that the ego is absolutely sovereign. The only time its boundaries might soften are when we are in love. Freud thought that this sense of interconnection between lovers was only an illusion but not pathological.[4] Today, people are more individualistic than perhaps at any point in the history of our species. Chil-dren are often urged to 'believe in themselves', and there is almost a religious fervour in statements such as 'You can be what you want, if you just try hard enough' (and perhaps the sheer number of applicants for talent shows such as *The X Factor*, regardless of their physiological ability to hold a tune, reflects this). The extreme individuality of our culture seems to shun external measures of self-worth, instead raising up our own self-assessment as king.[5] Sovereignty of the individual is seen as a human right to protect and cherish.

There are pros and cons to different worldviews. Excessive value of family honour with no regard for individual rights leads to terrible phenomena such as honour killings. Human rights principles that recognise we should not undertake actions that harm other people are certainly worth cherishing, but it's all about balance – there are clear costs both to ourselves and the planet arising from excessive individualism. Perhaps the culture of extreme individuality will not stick around forever; like a pendulum swung too far one way, hopefully it will soon return to settle at a balanced equilibrium.

The evolution of culture is sometimes viewed in a parallel way to genetic inheritance. New cultural innovations are passed down vertically through generations and also between peers of the same generation. These innovations (or 'memes' as they are referred to, as counterparts to their biological cousins – genes) bring benefits to their bearer and the close social group of which they are part, making them more likely to persist and spread. In his book *Wired for Culture,* evolutionary biologist Mark Pagel describes how biological and cultural inheritance are not distinct processes, but rather our human brains have evolved to make use of cultural memes to benefit us and our families.[6] Cultures of arts, music, religion, he argues, are memes which have helped individuals prosper. What's more, our genes biologically programme into us a predisposition to rapidly learn languages and culture from others. So from a slow languishing prelude for millions of years in which evolution was determined only by natural selection of genes, in very recent evolutionary time the tempo has suddenly picked up and the coevolution of genes and memes are inextricably linked in a close quick-step.

A similar interaction between our biological genes and cultural memes is seen in the perception of ourselves as distinct individuals. It is highly likely that having a distinct sense of self evolved because it brought adaptive benefits, particularly in a social context. As previously discussed, it would have benefited individuals to have a coherent sense of identity, allowing effective memory recall, problem-solving skills and an understanding of what others are thinking. Additionally, it would benefit small groups of humans to have members that saw themselves as individuals, allowing them to understand each other effectively and adopt specialist roles. For a group to thrive there is a fine balance to be struck between members being partly individualistic versus taking this position to an extreme level of selfishness and losing all collective identity needed for the group to function. This balance of individualism versus collective identity may well have been honed by natural selection. Groups that were full of extreme individualists, only out for their own benefit, would be forever plagued by cheats who thwarted any collective actions. These cheats would carry out selfish behaviours like stealing food that had been stored by the group, attempting to mate with other group-members' partners and avoiding the danger of predators by always hanging at the back of hunting parties and skimping on work to be done. They would cunningly avoid any altruistic acts that might benefit the group but put themselves at higher risk. However, because groups cannot function well with such members, these cheats would be punished by other group members – they may have been physically beaten or driven away, or put on such a low peg on the social ladder they failed to find willing mates. We see such punishment of cheating in monkeys and apes, where dominant individuals punish subordinates who break the norms of reciprocal cooperation.[7]

At the other end of the spectrum, groups in which the members had no sense of individual identity at all would collapse, because the members themselves simply wouldn't be able to function as self-sustaining agents. As we have considered, an individual identity is needed for the most basic tasks, and especially to function effectively in complex social environments. So, natural selection may have culled in such a way resulting in humans with a certain balance of individuality and collective identity (often called balancing selection), with brains physiologically fine-tuned by evolution to hold such a perspective. To make the analogy slightly more sophisticated, it would be a *double-edged* blade of natural selection: individual and group selection as the two agents crafting the evolution of individuality. If individuals were too selfish, they would face direct costs such as being physically beaten or failing to mate (individual selection operating through reduced reproductive success), but they may have also made the group itself less successful overall (meaning that it was outcompeted by other, more cooperative groups, who then became dominant). Cultural evolution would then take the reins and reinforce the balancing evolution from natural selection. For example, group members would be organised under an agreed set of norms and rules for how individuals ought to relate to each other, thereby allowing further optimisation of individual and group benefits.

It may seem a remarkable marriage between biology and sociology, but there is a further nuance in this evolutionary waltz of genes and memes. We know that not all genes that pass through populations of plants and animals actually benefit their bearers. There is a subset of genes, such as those carried by viruses, that can insert themselves into the host genome and impose a strong cost to the host individual, in some cases causing severe illnesses (for example, HIV virus causing AIDS). We heard about these manipulative viruses in Chapter 4. In a similar way, there can be 'viruses of the mind', which are cultural memes providing no benefit at all, or even a cost, to the minds they occupy. Some of these viral memes are relatively harmless, such as cheesy radio jingles that get stuck on repeat in your head and you just can't seem to shake. Others can be more pernicious, like self-harming, or xenophobic and homophobic attitudes. As in the contagious suicide trends in Manitoba described earlier, these destructive memes can rapidly spread through social networks under certain conditions.

Because not everything that evolves is always beneficial to the bearer, we might enquire whether the level of individuality modern humans exhibit is optimal. Is the level of individuality we currently possess ideal for our wellbeing and survival, and for the social groups with which we share a collective fate? One reason to suspect that average levels of individuality might not be ideal comes from the observation that a number of other human traits seem to have become sub-optimal (in other words potentially maladaptive) in the modern world. For example, the epidemic of obesity we face in the developed world – with over one quarter of the world's current population overweight or obese – stems partly from the innate desire to crave and seek out sweet and fatty foods. In prehistoric times, such urges would have been adaptive, ensuring we obtained sufficient energy-rich food sources in the transient windows they were available. In the modern world, however, where fats and sugars are cheap and superabundant in our supermarkets, food cravings lead to obesity, with its significant associated health costs. If the physiological costs of a genetically based trait are very high then natural selection will limit the spread of the trait through populations – through higher death rates there will be a reduced number of individuals holding those genes and lower transmission to future generations. But, in our modern societies, the keen blade of natural selection is dulled on the shield of our medical healthcare systems. Mortality rates are no longer much higher for individuals with such genes, at least not so much that the number of offspring and the transmission of genes are significantly affected.[8] Another example of this is our eyesight. In early humans, very poor eyesight would be hugely detrimental, limiting an individual's ability to gather food, avoid predators and to successfully raise children. Therefore, although there would be some variation in visual acuity between people, it wouldn't have been nearly as large as it is today. Nowadays, the selective pressures are very different. We rarely have to avoid predators or forage painstakingly to gather our own food. In fact, the remaining major survival threats we face related to visual acuity are probably our ability to avoid traffic accidents.[9] Even this threat is on the wane as road management systems improve in many cities across the world, reducing the frequency of vehicle–human collisions and the strength of natural selection. We have also developed techno-logical fixes, such as glasses and contact lenses, to correct for poor vision. By these means, we protect the individuals that suffer from these afflictions, which is absolutely the right thing to do, but it comes at a collective cost of increasing the relative frequency and intensity of such conditions in the population as a whole.

What about the trait of selfishness and the control of its in-tensity in a human population? As discussed above, if a person shows excessive individuation to the point of selfishness then they will not function well in a group, which will be to their own and the group's detriment. But what if cultural changes are able to act as a shield against the sword of natural selec-tion, as in the case for visual defects? One such cultural change could be the formation of larger groups in human societies. Being a cheat in a small group is very bad for the group and, as a consequence, there is strong incentive for cheats to be weeded out and punished. It is also easier to identify cheats in a small group, so they are less likely to remain undetected. Sci-entific studies across species from howler monkeys, wild dogs, to lemurs and wolves, support this prediction and find that cheating is less feasible for individuals in smaller groups and, consequently profitable only for individuals in larger groups.[10] So back in our evolutionary history, when we roamed across the plains of Africa in small bands of humans, or lived in small closely knit townships, there would have been strong controls on the frequency of social-cheats through punishment and threat of social exclusion. However, over the course of our human evolution, social group size has continually increased: from family groups, to tribes, to settlements, to countries and to international groups. Where humans were once part of many small groups at the mercy of strong natural selection on individuals and potentially selection against entire groups, we are now part of one great interconnected society whose shared culture (healthcare, welfare, social norms) protect us from harm. But what if this comes at the cost of harbouring and even nurturing pathological elements? In larger groups, extreme levels of individuality and selfishness are no longer effectively punished. It may well be that the natural evolutionary controls on the level of individuality expressed by humans have been lifted so it has now become a runaway process, ultimately leading to widespread pathological behaviours. Of course, there will come a tipping point for any group when a large proportion of members become selfish cheaters and the whole group collapses; in a globalised society that single tipping point could be catastrophic.

Let's hope we can develop some foresight as a species regarding this imminent danger and guide ourselves along a wiser path. Perhaps we may be rescued if the excessively high levels of individuality begin to trigger some kind of direct costs to individuals, spurring them to self-correct their behaviour without the need for group selection on an epic level. This individual-cost hypothesis could explain how the epidemic of developed-world obesity might slowly be reversed.[11] Warnings to individuals such as the threat of developing diabetes, heart disease or any number of obesity-related ailments, if well communicated, can spur intelligent people to change their behaviour and override the maladaptive cravings for junk food their genes impose upon them. So what are the personal costs of excessive individualism? Just find the most selfish or most lonely person you know and ask them how satisfied they are with their life . . .

13

Positive connections will improve your health

In absolute silence, Abe Kenzo sits in his armchair staring out of the small window. The cherry blossom drifts aimlessly off the trees and the breeze causes a gentle swaying of the curtains, which is the only movement in the small room. Abe sits like a statue, the lines on his face drawn in a thoughtful frown, his lips parted as if about to ask a question. The same expression has rested on his motionless face for many hours. The only difference now is the tautness of his skin and the slowly decaying flesh beneath.

Abe has been sat in this chair for just over three months. He is a sad case of *kodokushi* (a 'lonely death'), the name given to Japanese people, mainly single older men, who have died alone in their homes. Isolated from all social ties, these people often remain undiscovered for many weeks. And their numbers are increasing, trebling from the early 1980s to '90s, then doubling again in the next decade. One estimate suggests that 32,000 elderly people nationwide died alone in 2009, with 4.5 per cent of funerals in the same decade involving instances of *kodokushi*.[1]

Loneliness and social reclusiveness in Japan is not restricted to the older generation. A phenomenon known as *hikikomori* describes Japanese, mainly adolescents, who withdraw from social life and refuse to leave their houses, in some cases for decades. Japan government figures in 2010 estimated there to be 700,000 *hikikomori*, with 1.5 million borderline cases (one of the official criteria for *hikikomori* is to be confined within one's home for over six months).[2]

Japan is not the only country to be experiencing such problems. Similar extreme social withdrawal has been documented across many other countries, and there is plenty of evidence of less extreme, but widespread, cases of social isolation. In the UK, for example, London has been dubbed the 'loneliness capital of the world' following a spate of recent surveys revealing a worrying epidemic. Exact figures vary depending on the survey and location, but a poll of over 2,000 UK adults in 2014 found that 68 per cent said they felt lonely sometimes, regularly or often.[3] The proportion was highest (83 per cent) in younger adults, those aged eighteen to thirty-four. Within this age range, the most lonely were twenty to twenty-nine-year-olds, with one-third describing themselves as lonely people. A separate poll conducted in 2011–12 found that 11 per cent of UK people feel lonely 'all, most, or more than half of the time'.[4]

This worrying phenomenon, which seems to be on the rise globally, has significant physical and mental health impacts. Social isolation and loneliness are implicated in a range of diseases, from high blood pressure to eating disorders, alcoholism and dementia. The negative health effect is estimated to be on a par to smoking fifteen cigarettes a day, and twice as deadly as obesity.[5] In the UK, the number of young people admitted to hospital for self-harm has risen by almost 70 per cent in a decade, while the number of young patients with eating disorders has doubled in three years.

Why is this happening and how can we stop it? There are many factors tied up with our modern lifestyle habits that drive increasing social isolation. How and where we work is important, with modern jobs often involving long commuting or with frequent house moves to follow careers, meaning that there is less embedding within a local community.[6] Only 58 per cent of Britons feel they know people in their community well.[7] The time pressures created by working a long way from home mean that many people eat meals alone, a time when social interaction would traditionally occur. For example, a recent poll found that 40 per cent of Britons say they eat breakfast alone most days of the week, while 30 per cent say they eat dinner alone more often than not.[8] What a sad situation this is, and even worse that a large proportion of the rest who do eat dinner together possibly do so in front of the TV.

How we spend our leisure time is important too. If not watching TV, many people relax by playing computer games or interacting with their mobile phones, but these screen-based activities come at the expense of direct and meaningful interactions with other people. The advent of social media has exacerbated problems markedly. While we have become more connected digitally, this has come at the cost of real face-to-face interactions, with people spending around two hours a day on average using social media according to many polls. Just the sheer time spent online rather than in physical social interactions is problematic, yet there is also increasing evidence that social media can cause anxiety and social isolation, especially among children. It is perhaps not surprising that social media can cause anxiety – as social animals, we evolved to be sensitive, and efficiently track, others' opinions of us. When we are in face-to-face contact with a handful of people, we can read physical cues to assess how we are being interpreted. In contrast, online, we are potentially engaged with thousands of others, with very little feedback on how they are interpreting us, leading to an inevitable sense of anxiety and worry. Add to that the problems with online bullying and we can see how social media could be a significant contributory factor in the rapid rise in childhood anxiety and self-harm in recent years.

Many psychologists believe the culture of extreme individuality in which we raise our children leads to anxiety and mental health problems. At a young age, school children are strongly encouraged to compete against each other through standardised tests. The winners may thrive (if they are not burnt out with the extensive preparations for tests), but what about the rest who are consistently labelled as underachievers? As mentioned, an additional compounding factor affecting mental health in children may be the lack of access to nature. It is becoming evident that green space and biodiversity can have psychologically restorative properties, reducing anxiety and depression disorders.[9] Yet, in the modern world, increased urbanisation and modern routines lead to many kids playing indoors and children with far less opportunity to spend time connecting with nature.

These well-documented health impacts of a disconnection

between people and with nature are a pathology in the true sense of the word. The word 'pathology' stems from the Greek term *logos*, meaning establishing an intelligent order of the universe, and *pathos*, meaning a disease caused by imbalance. In the modern world, our understanding of the order of the universe is increasingly limited by the illusion that we are discrete, isolated entities, and this is causing disease-like symptoms – damaging both our personal wellbeing and the wider environment that support us. In philosophy, the term *individuation* is used to describe the process by which an object, or ourselves, becomes distinguished as distinct from everything else. Putting these terms together, a term to describe our modern malady is *individuation pathology* – the set of disease-like symptoms caused by a fundamental miscomprehension of our connectedness to the rest of the world.

So how do we treat this individuation pathology? Fortunately, there are examples throughout history where authors have had insights into this problem. As early as 200–500 BC the ancient Hindu text, the *Bhagavad Gita* taught that, fundamentally, the sorrows of human life are caused by humans misconceiving their own nature.[10] Buddhism holds this as a strong central tenet too, with the Buddha teaching that we can be released from suffering only if we let go of our desires by realising that the self is an illusion. Similar sentiments are repeated in many secular texts, as well as other religions, for example in the Bible: 'There is neither Jew nor Gentile, neither slave nor free, nor is there male and female, for you are all one in Christ Jesus.'[11]

Yet, with all this accumulated insight, today we are

unfortunately no closer to seeing through the illusion of our self-absorption and resolving our individuation pathology. In fact, as a society we may be further away than ever, given the increasing occurrence of loneliness and nature-deficit-related disorders in the modern world. How is this possible, when our shared human knowledge should be accumulating over time and revealing solutions to our problems? We have all the insights of these ancient religions plus the rapidly growing knowledge of modern science – where did it all go wrong?

Of course, wisdom is not just about having the knowledge in a library, or reading and understanding it only in theory. Theoretical knowledge is very different to knowing something through direct experience and integrating it into our deeply held inner beliefs about how the world works, and so it is for the wisdom needed to overcome the self delusion. All psycho-logical diseases have a physical representation in the brain. These result from thought patterns and behaviours hard-wired into our neural connectomes through a combination of gen-etic determination and from our lifetime experiences. As we encountered earlier, neurons in our brains that 'fire together wire together', and so certain well-used thought patterns and deeply held belief structures can be very hard to escape from because they become literally hard-wired in our brains. We face a significant task in shifting electrical brain waves from these well-trodden tracks to new patterns – from the illusory belief in independence to a truer comprehension of intercon-nectedness to the world and other people around us.

Think about that moment when you wake up in the morn-ing and your mental computer jumps out of sleep mode and begins the waking 'self' program. We conceive of ourselves as distinct from the world outside our physical bodies, using layer upon layer of mental abstraction, building great false castles of thought that separate 'us' from the world. Yet the science now shows these belief structures are essentially false. Our brain has evolved so that it efficiently runs a program (let's call it 'IllusionofIndependentSelf V1.1'), which places us in a mental cockpit of illusory autonomous control of a discrete self. In terms of neural networks, this program runs on well-worn neural pathways, like deep ruts in a road that a cart travelling along would struggle to leave.

This may explain why knowing a truth *theoretically* doesn't do much to free us from it. We need to learn a new way of thinking and *practise* it like any skill. Explaining to someone exactly how the muscles in an Olympic archer's arm function to fire an arrow seventy metres into a target doesn't get you very far in carrying out the same task yourself. Instead, repeated training is needed to develop the neural pathways and learn the skill. The same is true for our perception of the world – we have the opportunity to cast away individuation pathology and see things accurately, but it is unlikely to be an instantaneous 'Eureka!' moment of understanding. It is likely to take hard work and practice to develop a new configuration of neurons in our brain. As Henry David Thoreau wrote in *Walden*: 'we are enabled to apprehend at all what is sublime and noble only by the perpetual instilling and drenching of the reality that surrounds us'.

So what is the best way to perpetually drench ourselves in this reality? The theoretical physicist David Bohm had an extreme suggestion: we need to change the everyday language we use. Along with his contributions to quantum physics, Bohm was deeply interested in questions of neuroscience and psychology, because he believed that ecological, social and psychological problems are caused by disconnectedness and fragmentation, which resonates with the disorder labelled here as 'individuation pathology'. According to Bohm, because we are steeped in our language, it conditions and reinforces the way we think. This fits with observations, described in Chapter 11, that people from Western cultures, who tend to perceive the world in a more fragmented way, use nouns more often than verbs. Using nouns more often reflects, and perhaps further reinforces, their belief of existing in an abstract world comprised of many discrete entities.[12] Bohm's idea was that nouns are fundamentally false, because they signify unchanging, isolated entities, when in fact everything in the world is in a continual state of flux and connected intimately to everything else, something that his work in quantum physics had revealed to him. Instead, he suggested we should replace nouns with verbs because these more accurately reflect the dynamic nature of the world. So rather than referring to writing parchment as 'paper' for example, we should use a word like 'papering', which recognises that the paper is in a continual state of flux, losing molecules to the atmosphere and slowly undergoing an unstoppable process of gradual degradation away from a paper-like state. This conceptualisation reflects the sentiments of other deep thinkers. For example, the Buddhist monk and prolific author, Thich Nhat Hanh, wrote:

This piece of paper I am holding in my hand is something that exists right now. Can we establish a time and place of birth for this paper? That is very difficult to establish, impossible actually, because before it manifested as a piece of paper, it was already here in the form of a tree, of the sun, of a cloud. Without the sun, without the rain, the trees would not have lived, and there would have been no piece of paper. When I touch this piece of paper, I touch the sun. When I touch this piece of paper, I also touch the clouds. There is a cloud floating in this piece of paper. You do not have to be a poet to see it. If I were able to separate the cloud from the piece of paper, the paper would not exist anymore.[13]

These are poetic words that may strike a chord within us as reflecting a deep truth; yet, if you feel that incorporating this kind of conceptualisation into our everyday language, for every noun that we speak sounds completely unworkable, then I would probably agree with you. Suffice to say, Bohm's suggestion of transforming our human language has not been broadly taken up. Imagine, say, a farm owner discussing with a contractor a job that needed doing on the farm. She says, 'George, can you mow the field today please' – a concise and to the point request. Now, the farmer may indeed conceive of the 'field' in an enlightened sense, as reflecting a dynamic intersection of many different ecological and socioeconomic phenomena: in addition to the grass and its value as hay, she may be aware of the changing balance of nutrients in the field and the state of the microorganisms that are the powerhouses of healthy soil; she may be cognisant of the bees that feed on flowers in the field and pollinate her nearby apple orchard and field beans; she may also be aware of mammals and birds that at various times of year use the field as habitat for shelter and resources, and the long cultural history of the field managed in different ways by many others before her. Yet, if the farmer tries to capture all this dynamic interconnectedness in her request to George, it would make for a very long conversation, and it would actually hinder getting the job done in a timely way.

The hundreds of human languages that exist vary greatly from each other, but at the basic level of communication, they are broadly similar. Most often, using sounds, aided by gestures, we communicate short salient messages that facilitate necessary tasks. Thus, succinctness and clarity are important factors driving language evolution. Some redundancy does occur, which can function in reducing error rates in the transmission of the message, but there is certainly not excessive redundancy. So, although nouns may be abstract concepts that are ultimately inaccurate in reflecting the deeper interconnectedness of objects in the world, as tools for concise language they are a pretty useful shorthand and, apologies to Bohm, they are probably here to stay. We will have to find other ways to 'drench ourselves' and engage in the deeper reality of the world that don't involve compromising our ability to communicate and get on with necessary tasks in our lives.

This point can be articulated more generally: although we are aiming to develop ways to overcome our individuation pathology (and so reduce personal loneliness and other associated maladies), we need to bear in mind that the sense of an independent self evolved in the first place because it brought us benefits. Therefore, we don't want to abolish the self *at any cost;* for example, by adopting a verbose and dysfunctional language, or by taking excessive doses of psychedelic drugs. The Romantic poet William Blake once enthused: 'If the doors of perception were cleansed, everything would appear to man as infinite!' A beautiful thought, to which the author Rushkoff pragmatically responded: 'We would be lost in that endless, limitless expanse!' Rather than a wholescale transformation, risking becoming lost in a limitless expanse without an anchor, we need to 'tweak' our worldview and redress the imbalance where the sense of self has become overly extreme and maladaptive. So, what are some other options to achieve this more nuanced realignment of the self?

One possibility is to train ourselves in a different way of perceiving the world. Many ancient Eastern traditions were progressive in this regard, laying out principles and methods to train the mind to view the world in a holistic way, which recognises the intricate connections between objects and phenomenon. Their fundamental tenets hold that independence is an illusion that causes unhappiness. For example, in Hinduism: 'One who sees everything as nothing but the Self, and the Self in everything one sees, such a seer withdraws from nothing. For the enlightened, all that exists is nothing but the Self, so how could any suffering or delusion continue for those who know this oneness?'[14] Such sentiments are sometimes reflected in Christian texts. For example, Meister Eckhart: 'The eye with which I see God is the same with which God sees me. My eye and God's eye is one eye, and one sight, and one knowledge, and one love.' Yet, it is religions such as Zen and

Buddhism that have laid out the clearest instructions to practically achieving such changes in perception. They describe ways to change our thought patterns, including meditation.

There is strong scientific evidence underpinning these transformations of our thought patterns. Behavioural psychology studies have shown how our perception of the world is flexible. Tests designed to assess the extent to which we perceive the world as a set of independent discrete objects, versus interconnected phenomena, are influenced by the prior state of the participant, in what are known as 'priming effects'. For example, when asked to fill out a survey ascertaining the importance of individualist or collectivist attitudes, participants' responses differed if they were asked to read a short story beforehand. Simply reading the story could prompt changes from 'isolated' to 'interrelated self-constructs' in the participants.[15] Similar priming effects also result in altered performances in the visual field dependency tests mentioned earlier, which aim to assess the degree to which participants can detect embedded figures from a background scene – ascertaining the degree to which they perceive and categorise the world into distinct isolated parts.

These experimentally induced changes in perception are in most cases quite temporary. Our perception is resilient to a degree and returns to a fairly stable level (which partly forms the basis of our personalities); but that is not to say substantial long-term changes cannot occur, after all, people's personalities do change over time. Training thoughts or actions involves repeating neural patterns of activation, leading to physiological changes in neural networks which then make such patterns of activation more likely. Recent advances in neuroscience now allow us to map these changes in neural connectivity as training occurs. This provides a firm understanding of how repeated practice in alternative ways of perceiving and thinking can cause long-term changes in the brain. Therefore, perhaps some kind of meditation approach which focuses on thoughts and feelings of interconnectedness, done correctly, could lead to the 'tweak' in our perception that we seek. Indeed, ECG scans of experienced meditators do show structural differences in the way the brain operates, with many meditators showing positive traits such as lower stress levels in response to stressful events and increased capacity for empathy with others. Of course, this works the other way too – we can train ourselves in bad habits. Our modern culture, from media advertising to our formal education system, all seem to emphasise extreme individuality, and these are arguably providing an overwhelming set of stimuli which are training our minds to increasingly perceive the world in an individuated way. We can be 'drenched' in negative environments less conducive to self-transformation.

Even if many hours of mediation can provide a much-needed 'cure' for individuation pathology, we don't necessarily have the time or inclination. Fortunately, there are other ways to immerse ourselves in alternative modes of thinking that may help to banish the illusion of independence. One is greater engagement with nature. The complexity and beauty of the natural world is a great way to draw ourselves out of a narrow self-centred worldview, as many nature lovers will attest. Another is spending face-to-face time with friends and family, which addresses our deep biological needs as social animals.

There is scientific evidence supporting these views as well. The New Economics Foundation, a UK think tank, synthesised the latest research and came up with a short, easy-to-do list of activities that promote happiness and wellbeing (in a similar vein to the healthy 'five-a-day' recommended for increasing fruit and vegetable consumption). The recommendations were as follows: *1. Connect with other people, 2. Take notice/be mindful, 3. Be active, 4. Learn something new, 5. Give and show kindness.* It is amazing how many of these depend upon recognising and appreciating our connectedness to the world around us. The first, *Connect*, involves engaging and sharing with others; the second, *Take Notice*, involves stepping away from an abstract conception of the world with us at the centre and perceiving more carefully the true reality of the world around us (in other words stilling the mental program of self-delusion and putting 'IllusionofIndependentSelf V1.1' into sleep mode). Number three, *Be Active*, means exercise, which we might not think of as having anything to do with connectedness. However, during and just after exercise, have you ever noticed how you are drawn out of your mind to feel a more rooted sense of groundedness? The Hungarian psychologist Mihály Csíkszentmihályi first described the sense of 'flow', the feeling of energised and focused absorption that can occur when people engage in a physical activity that requires some concentration and skill. The state has now been studied across many disciplines. Along with factors such as intense awareness of the activity and alteration of the subjective experience of time, one of the primary correlatives of the experience of flow is a loss of self-consciousness, a dissolving of our sense of discrete individuality.[16] The fourth recommendation for mental wellbeing, *Learn*, involves better understanding of different aspects of the world and, crucially, how these interrelate to each other. Learning is about connecting and synthesising knowledge of the world. Incidentally, it also involves increases in the connectivity of neural networks in the brain, our connectome. The last recommendation, *Giving*, whether it is through money or our time to others, involves challenging our sense of possessiveness and breaking down self-defensive barriers.

So it appears that there are a number of options that can help to reduce individuation pathology, and perhaps taking some of these steps will tackle the epidemic of loneliness and associated mental and physical health problems in the modern world.

14

You can't blame someone who doesn't exist

We need not shrink from this comparison of small things with the great; for does not science tell us that the highest striving is after the ascertainment of activity which shall bind the smallest things with the greatest? In natural science, I have understood, there is nothing petty to the mind that has a large vision of relations, and to which every single object suggests a vast sum of conditions. It is surely the same with the observation of human life.   George Eliot, *The Mill on the Floss*

'It was after much thought that I decided to kill my wife, Kathy, tonight after I pick her up from the Telephone company. I love her dearly, and she has been as fine a wife to me as any man could ever hope to have.'[1] Thus wrote twenty-five-year-old Charles Whitman on the evening of 31 July 1966, shortly before plunging a knife three times into his wife's heart while she slept. He had also killed his mother similarly before proceeding to climb the University of Texas tower with seven guns and ammunition in a bag. At the top, Charles killed a receptionist then shot two families who were behind him on the stairs. He then took aim from the observation platform and shot a pregnant woman on the ground below. As her husband ran to assist his dying wife, Charles shot him too. In a shooting spree, he then killed a further sixteen people and wounded around twice as many more.

As a judge, what sentence would you give to such a cold and calculating killer? If he stood before you now, would you loudly denounce Charles Whitman as 'evil' and rule the harshest punishment upon him? As it was, Whitman was shot and killed by police and a heroic volunteer at the top of the tower; but, had he stood trial, the case would have been very interesting, highlighting the complexity of attributing blame. No doubt, many would have called for the most severe punishment for Whitman – death by lethal injection, or a lifetime in prison. But would there have been others who tried to tease out the circumstances that led to an intelligent young man acting so destructively?

Whitman himself recognised his behaviours and tendencies as unusual. In his diary, he wrote: 'I don't really understand myself these days. I am supposed to be an average reasonable and intelligent young man. However, lately (I can't recall when it started) I have been a victim of many unusual and irrational thoughts. These thoughts constantly recur, and it requires a tremendous amount of effort to concentrate on useful and progressive tasks.' He took himself to a psychiatrist and talked about how he was experiencing overwhelming periods of hostility with little provocation, and admitted to thoughts of 'going up on the tower with a deer rifle and shooting people'.[2] In his final letter, Whitman asked for an autopsy to be performed to see if there was anything physically wrong with his brain that could explain his persistent violent thoughts and the painful headaches he was frequently experiencing.

Upon performing the autopsy, medical examiners indeed found a tumour lodged inside his brain. It was compressing the amygdala, an area of brain circuitry involved in regulation of fear and aggression. Earlier studies in monkeys had shown that damage to the amygdala leads to a range of responses, including lack of fear, blunting of emotion, and overreaction. Female monkeys with amygdala damage often neglect or physically abuse their infants.[3] Could this explain the unusually powerful and out-of-place emotions that Whitman was experiencing? A commission of neurosurgeons, psychiatrists, pathologists and psychologists convened by the Texas state governor concluded that, indeed, it could conceivably have contributed to his inability to control his emotions and actions.

A number of other medical cases now show clearly how brain tumours can cause their sufferers to experience violent, or sexually deviant, tendencies. These behaviours, in some cases, instantly disappear when the tumour is surgically removed and reappear if it starts to grow again.

There are other physical origins of behaviour change in the brain that are less apparent than tumours. For example, traumatic experiences during early childhood, a period when our brains are developing rapidly and are especially sensitive, can lead to long-term structural differences in the brain, fundamentally affecting the way it operates. Experiments in monkeys (that wouldn't pass ethical approval nowadays) show how infants, well fed and watered but kept in isolation for longer than the first six months of their lives, developed abnormal behaviours, rocking back and forth, biting themselves and attacking their own babies when they became mothers. Clearly, there is a critical period during which healthy social interaction is essential for the development of a correctly functioning brain. Social experiences during this period shape the neural connections in the brain, leading to permanent personality changes later in life. Similar results of social isolation have been found in other animals ranging from rabbits to pigs, rats, mice and even fruit flies.[4] The socially dependent part of our brain seems to have a deep evolutionary history. Such isolation experiments have not been permitted in humans, although there have (unfortunately) been historical episodes of social deprivation that have led to insights into their impacts in our species.

In the 1990s, the world was outraged as it was revealed how hundreds of Romanian orphans were being kept in cramped, filthy orphanages with poor nourishment and almost no social interaction with caregivers. Some of these children were subsequently rehomed in the UK and the psychiatrist Sir Michael Rutter studied the progress of their recovery.[5] Sadly, on arrival many of the babies showed strong signs of mental retardation and were physically stunted in their growth. Many of these symptoms disappeared over subsequent years, and the fostered orphans caught up to normal children to some extent, in IQ tests and other similar measures. However, following these children into their childhood and early teens, Rutter's team found they had problems with hyperactivity and in forming relationships. These problems were most marked in those who had spent the longest time in the orphanages during what the researchers found to be a critical developmental period. Experiences that happened to them during early years, which were completely out of their control, produced a physical manifestation in their brains, for example, limiting the inhibitory action of higher cortical areas, which strongly influenced their subsequent behaviours.

This sad example resulted from social deprivation combined with malnutrition during the early years of life. Outcomes can be worse still if young children are physically abused. There is a worrying abundance of evidence showing how families with higher levels of physical abuse often raise children who are unable to control their tempers and become physically violent themselves. This is particularly true if these children are born with certain combinations of genes, making them more likely to suffer from antisocial behaviour when raised under cruel environments.

The correlation between certain genes and anti-social behaviour is clearly recognised. For example, a 2011 article in the magazine *Atlantic* explains:

if you are a carrier of a particular set of genes, the probability that you will commit a violent crime is four times as high as it would be if you lacked those genes. You're three times as likely to commit robbery, five times as likely to commit aggravated assault, eight times as likely to be arrested for murder, and thirteen times as likely to be arrested for a sexual offense. The overwhelming majority of prisoners carry these genes; 98.1 percent of death-row inmates do.[6]

However, being born with such genes does not seal your fate making such outcomes inevitable ('genetic determinism'), rather it slightly increases your chances of developing anti-social behaviour. Plenty of people have the genes and suffer no problems. Antisocial behaviour arises through a combination of *both* certain genes and certain life experiences. For example, researchers have found that children that victimise others were more likely to do so when they were maltreated as children *and* had an abnormality in the gene encoding a specific neurotransmitter-metabolising enzyme.[7] So it is not so much genetic determinism as 'nature-*and*-nurture' determinism.

This past influence of both genes and experience ('nature and nurture') in determining behaviours raises questions of quite how culpable someone should be for their actions, and of course, it has big implications for legal cases. As science gradually reveals how violent criminality is often explained by disruption of the neuronal circuits that modulate aggression, lawyers increasingly refer to the biological basis of behaviour in defence of their clients' actions. This doesn't necessarily change the fact that we will often need to lock up perpetrators in order to keep other people in society safe, but it does change the way we view these people as criminals who should be punished. In fact, the whole basis of court rulings as punishments starts to fall away. If someone's actions are in large part determined by their genes and events imposed upon them of which they had no control, can we really *blame* them for these actions?

The issue of whether we can validly lay blame on someone depends to some extent on our belief in the existence of free will. If we do have free will then we are fully responsible for the actions we take and we can be blamed for them. However, many scientists (and deep thinkers long before the advent of modern science) are of the opinion that such free will may be an illusion. As neuroscientist Bruce Hood explains:

> Any choices that a person makes must be the culmination of the interaction of a multitude of hidden factors that range from genetic inheritance, life experiences, current circumstances and planned goals ... We are not aware of these influences because they are unconscious and so we feel that the decision has been arrived at independently.[8]

The seventeenth-century philosopher Spinoza recognised this too: 'Men are mistaken in thinking themselves free; their opinion is made up of consciousness of their own actions, and ignorance of the causes by which they are determined.'[9] There has been resistance against this view, because of the fear that it encourages a fatalistic perspective and could prompt people to engage in criminal behaviour in the knowledge they will not be held responsible. This fear could be misplaced, however, because taking a stance that free will does not exist can still be perfectly consistent with the belief that people should act responsibly and should be *liable* for their actions. So perpetrators may still need to have corrective measures taken against them, in order to protect society and prevent recurrence of the behaviours, but the door is now open for a more compassionate treatment of such miserable wretches. Instead, argues Yuval Noah Harari: 'Our judicial systems largely try to sweep such inconvenient discoveries [regarding limited free will] under the carpet. But in all frankness, how long can we maintain a wall separating the department of biology from the departments of law and political science?'

It is not just our legal systems that are amiss here. Our institutions tend to reflect the mindsets of the people who create them. In our day to day life, we often blame others for their minor trespasses against us. We fail to recognise the multiple causes, acting beyond and through that person, that may have led to those events. Human brains are generally not good at understanding the true complexity of causality in the world, but rather tend to focus on a single salient cause. The evolutionary psychologist John Tooby suggests this is an evolved aspect of our cognitive machinery. We have evolved heuristic – rule of thumb – approaches to understanding causality that identify each event as having only a single obvious cause. He describes how we typically trace a causal chain backwards to end at another person, but rarely beyond. This allows us to 'punitively motivate others to avoid causing outcomes we don't like (or to incentivise outcomes we do like)'.[10] Blame, he suggests, is an evolved cognitive tool to manage both the complexity of the real world and our social interactions with each other. Ironically, when bad things are caused by us, we are more likely to attribute them to outside factors beyond our control. So there seems to be a cognitive bias at play, meaning we have one rule for us and a different rule for others.[11]

In reality, blaming something on another person is just a lazy shorthand. It is a failure to appreciate the ultimate causes of the person's behaviour, which may extend well beyond their control and even backwards to a time before they were born.

This failure to see the context of an event is more apparent in certain cultures, and particularly in our currently dominant Western culture, as we have already learnt. For example, in 1994 two University of Michigan researchers, Michael Morris and Kaiping Peng, carried out a review of newspaper stories from different countries that had reported on local crimes. They found American reporters often laid blame on the perpetrator (attributing the crime to their disposition), while Chinese reporters blamed situational factors.[12] These conclusions mirror those from controlled experiments, where participants from America and Korea were provided with information about a murderer. The American participants discounted over half of the information as being irrelevant to the crime believing it to provide unimportant context, while Koreans discounted only a third.[13] This shows how, with an extremely atomised 'Westernised' perception of world, we see individuals as increasingly blameworthy and ignore the context that can often help to explain the ultimate causes of human actions.

Although such a worldview is now dominant, it does not have to be this way, and in fact, as we have discussed, it has not been through much of our history. In ancient Eastern cultures, Buddhist scholars have long taken a holistic, systemic perspective on causality, recognising multiple causes to each event (the principle of *Pratītyasamutpāda*, or 'dependent arising'). Perhaps, as a consequence there is a greater focus on cultivating empathy and compassion in Buddhism. Meanwhile, other religions such as Christianity suggest we practise forgiveness when we are wronged – 'Forgive us our trespasses and those who trespass against us' teaches the Lord's Prayer. However, it can be hard to instantly forgive someone who, in cold calculation, has shot over forty-five people, as Charles Whitman did. What if one of your family members was killed? We might not *want* to forgive. In such circumstances, forgiveness can seem a superhuman feat. In contrast, trying to understand people in their situational context as part of a broader interconnected system can lead to a greater understanding of the possible causes of their actions, and may at least allow empathy – if we were in exactly the same position as them with the same history, we probably would have acted similarly.

In order to overcome our innate cognitive biases, we need to once again bring empathy and compassion into our culture and institutions. Such an attitude promotes understanding towards those who commit social wrongs and will help in finding ways to rehabilitate them where possible. In contrast, laying blame stigmatises perpetrators and can actually exacerbate problematic behaviour, causing a downward spiral. This can begin from a young age – for example, labelling a child as the naughty one in the family or classroom. Recent social research reveals that people respond strongly to subtle stereotyping and their personalities go on to develop in ways which reinforce such stereotypes. The words 'naughty child' are just a shorthand for a set of conditions in which a child displays behaviours deemed as socially unacceptable. But a naughty child is not an immutable entity, they are a young person who acts in that way for a variety of reasons, many which are outside of their control – perhaps they have been starved of appropriate attention and play, which facilitates socialisation. We need to be aware and ever cautious that the word 'naughty' is an abstract generalisation, just as the word 'criminal' is. We need these words for effective communication – we encountered earlier how we need such abstractions in our language to make it functional, but let us remember these are shorthand, and where they refer to people who we potentially stigmatise and punish, it is important to see through them and appreciate the root causes of behaviour, which may be well outside of their control.

Of course, what we consider here does not just apply to criminals, but to everyone we meet on a day-to-day basis. When we feel the urge to blame someone for minor trespasses, we might consider how the connections and causes are often well hidden, although to reveal these we need to take a holistic perspective and look beyond the individual. Understanding the science of our interconnectedness will help in promoting this systemic, holistic perspective too.

To give one last example of our tendency to thoughtlessly lay blame, consider the growing epidemic in obesity, which imposes large costs on the health service. There can be a temptation to stigmatise obese people – why don't they simply stop being greedy and eat less? Of course, we now know that eating disorders have complex psychological explanations, sometimes reflecting a traumatic history of the patient. In other cases, there may be complex physical causes. Recent research has suggested epigenetic effects that influence obesity, whereby the environment that mothers, or grandmothers, experience can alter their DNA, causing developmental differences in offspring and increasing the risk of obesity.[14] Changes in our gut microbiome have also been linked to weight gain, so can we really blame a sufferer of obesity for these kind of intergenerational and wider environmental causes?

Because of the complex pathways of causality extending in subtle ways beyond individuals, we should also be very careful of how our own actions affect others. Social-network theory shows how behaviours often spread up to three people away in social networks – we may be influencing people we don't even meet.[15] Also, because of the epigenetic effects described above, our behaviours and actions can affect our unborn children. Things like our diet and whether we choose to smoke or not can all have significant consequences on future generations. Therefore, adopting a perspective which looks beyond the individual can encourage a more compassionate approach in how we treat other people in day-to-day life, and also an increased awareness of our indirect effect on others, which may not be immediately apparent. The web of connections – of causes and consequences – reaches out far and wide in many directions from any event, or from any person. Understanding how causality can ripple outwards from immediate actions, leading to hidden consequences a long way away in space or time, may allow us to reflect more deeply on the impact of our actions, and this may, in turn, promote responsible actions. In a sense, compassion works both ways – it involves understanding the real root causes of the wrongs done to us, and also in recognising and preventing the hidden wrongs we do to others.

Changing our personal perspective to be aware of these extensive interconnections and develop a compassionate approach is also the first step in engendering wider change in our society. With regards to how we think about blame and punishment, it may transform our justice systems into more reflective and compassionate institutions, where we better unpick and understand what causes criminality. Thinking about our own actions, we may start to acknowledge and limit the negative impacts we have on other people and the natural world, paving the way to act in a more ecologically responsible way.

PART FOUR

OUR NETWORK IDENTITY

If for a moment we make way with our petty selves, wish no ill on anyone, apprehend no ill, cease to be but as a crystal which reflects a ray – what shall we not reflect! What a universe will appear crystalized and radiant around us.

Henry David Thoreau, Walden

I share my body with countless others,
My DNA is part of the web of life,
My brain is a product of my culture,
My heart is a product of humanity,
My emptiness is universal[1]

15

Three dimensions of interconnectedness

Since the cut between self and the natural world is
arbitrary, we can make it at the skin or we can take it
out as far as you like – to the deep oceans and distant
stars ... If psychology is the study of the subject, and
if the limits of this subject cannot be set, then psychology merges willy-nilly with ecology.   James Hillman[1]

In the late 2000s, a young American medical student, Amy
Proal, made a key advance in the understanding of a mysteri-
ous disease called chronic fatigue syndrome and it all began
by her going to bed – for nearly two whole years. Normally
a very active and motivated person, Amy was making good
progress in her degree when she started feeling unusually tired
and unwell. Her symptoms, which included severe headaches,
flu-like issues and muscle pain, worsened until she became
essentially bedridden. She was diagnosed with myalgic en-
cephalomyelitis (ME) – commonly known as chronic fatigue
syndrome – a disease whose cause was poorly understood.

Not one to take things lying down for long, Amy was determined to understand the science behind her affliction in

the hope she might find a cure. She sought out other people suffering from the disease, as well as those who studied it, and eventually met Professor Trevor Marshall, who was working on a maverick new treatment related to the human immune system. His theory was that runaway inflammation in the body, linked with a disrupted microbiome, was the cause of chronic fatigue syndrome.

Professor Marshall is a classic polymath: he once designed electronic music synthesisers, built the PA amplifiers used by the band AC-DC and has designed his own miniature surround-sound Hi-Fi system. When not tinkering with electrical circuit boards, he is also Director of the Autoimmunity Research Foundation and a world leader in understanding how microbiome disruption drives certain diseases. Chronic inflammatory diseases are routinely treated using drugs that suppress the immune response. The drugs lead to amelioration of some symptoms but do not resolve the disease and may subsequently encourage relapse over time. Instead, counter to the prevailing dogma, Professor Marshall's treatment was designed to boost the body's immune response, effectively kickstarting it into action. Such an approach can lead to an initial worsening of symptoms, as pathogens die-off and their toxins are cleared from the bloodstream. Therefore, he needed some brave participants to test the experimental treatment.

Having studied the scientific logic underpinning the proposed treatment, Amy volunteered to participate. Along with several other participants, she took a drug called olmesartan medoxomil that is standardly used to reduce blood pressure, but took it in higher doses and much more frequently than normally prescribed. The drug interacts with the Vitamin D receptor in the body cells – a receptor that is critical to normal immune function because it regulates over one thousand other genes, including those coding for antimicrobial substances. Immune responses present a serious challenge to pathogenic bacteria that want to colonise our body. As a consequence, some bacteria have evolved to interfere with our Vitamin D Receptor function, producing chemicals that mimic human ones to downregulate its functioning. This stealthy alteration of the immune defence system means bacteria effectively persist and thrive, leading to a chronic state of microbial imbalance in the body. Professor Marshall realised what was needed was a way to boost the action of the Vitamin D receptor to overcome this dampening by the bacteria, and that is what the drug olmesartan seems to do.

It takes time to rebalance your microbiome, to restore the complex interacting communities of bacteria and other microorganisms that underpin a healthy state, but after several months of the experimental treatment many of Amy's symptoms began to improve. The episode inspired her to steer her career towards better understanding the human microbiome so that she might harness new scientific discoveries to help others. Working closely with Professor Marshall and several other scientists, she researched further into the mechanism of diseases like chronic fatigue syndrome. Amy changed hats from patient to scientist and wrote up the experiment, explaining how the stimulation of the immune response is key to treating many autoimmune diseases.[2]

Understanding these interactions with the non-human

partners that comprise our microbiome – the interconnect-edness with our inner ecosystem – is essential for our health. When we don't take care of our human microbiome it can enter a state of imbalance ('dysbiosis'), where the composition of bacteria and other microorganisms in our body change semi-permanently, often with detrimental consequences. We now know that microbiome changes are responsible for problems like irritable bowel syndrome and contribute to the rise of obesity in the modern world. In recent years, there has been an explosion of scientific studies linking the microbiome to various pathologies, including gastric disorders, diabetes, auto-immune disease, asthma, and neurological conditions such as Parkinson's and multiple sclerosis,[3] although in some cases the direction of causality still remains unknown – an altered gut microbiome could be causing these conditions or might just be an additional symptom.

Because understanding these interconnections is so important to our health and happiness – something Amy Proal experienced directly – she has worked tirelessly to synthesise the science of how microorganisms influence nearly every aspect of our metabolic functioning; yet most medical schools still treat the body beyond the gut as sterile, despite the wealth of evidence on how the microbiome permeates most of our tissues. We also have a 'virome' – 30 billion or so virus particles – that travel throughout our bodies. This inner ecosystem mediates the effects of drugs we take, so it is essential that doctors and drug companies better understand these interactions to make medicines more effective.

We need to think about how new chemicals and technologies

might impact the microbiome. Thousands of new chemicals are registered every day, outpacing the ability of regulatory authorities to formally assess health risks.[4] They appear in diverse products from food, to paint, to shampoo and it is a chemical Russian roulette as to whether they impact our microbiome. New technologies on the horizon, such as nanotechnology, also pose a threat. To avoid exposing ourselves as guinea pigs to the potential risks of these new technologies, we need much greater cross-talk between scientists, policy makers, food and technology companies to make sure new products are safe – both for human cells and for the beneficial non-human partners in our bodies.

Through researchers like Amy Proal we are learning new ways to better manage our microbiomes to achieve healthier bodies and minds. Neurobiologists, microbiologists, immunologists and physicians all share part of the solution to this great puzzle. It is not an exaggeration to suggest that such understanding could extend healthy life expectancy considerably in the near future. There are great benefits to come from understanding the intimate connectedness with the multitude of partners in this communal vessel that we call a 'human' body.

Zoom out for a moment now, from thinking of your inner ecosystem to the outer ecosystems around you. Just as understanding the microbiome has clear health benefits, what about understanding our connectedness with the natural world outside our bodies? To learn more about this domain of outer connectedness we need to travel to rural Derbyshire, England, where we will meet an unlikely pioneer of the field – a man who used to design flat-pack furniture of the sort you might buy from IKEA.

Miles Richardson, Professor of Human Factors and Nature Connectedness at the University of Derby, is a researcher who works tirelessly to synthesise scientific knowledge about our human interconnectedness with the natural world around us.[5] He draws together research on how our relationship with nature affects everything from our mood and behaviour to our physical and mental health. In recent years, such psychological research on human connectedness with nature has exploded.

Miles did not always study this topic – he began his professional academic career in ergonomics, investigating how people relate to design tasks such as putting together flat-pack furniture. To counter a sedentary academic lifestyle, Miles began going on regular walks, taking notes about the nature he saw. After doing this every day for almost a year, he had enough material to document how his contemplative approach had changed his perception of the natural world and his place in it. He began to consider how his expertise of ergonomics could be extended to how people relate to the natural world, and it wasn't long before he steered a new career path as an environmental psychologist. Although there have been many historical and contemporary literary figures writing about such issues (Henry David Thoreau, Aldo Leopold and Rachel Carson to name just a few), it is only in recent years that a rigorous evidence-based approach has been used. New scientific approaches pull together measurements from questionnaires, geographic information systems, biophysical measurements (such as cortisol in saliva, a marker of stress) and even brain-scanning techniques, to help us understand why engaging with the natural world is so valuable.

From such research we now know that living close to and spending time in nature conveys a huge range of benefits. A greater psychological connectedness with nature helps to keep us mentally balanced and, in turn, motivates us to protect the natural environment that sustains us. In the face of increasing mental health crises in many countries these benefits of nature engagement have finally begun to appear on the political radar, with a growing interest in how to plan green space into the design of towns and cities (so-called green infrastructure). In 2017, an evidence statement produced by the UK government reviewed the health benefits arising from exposure to natural environments.[6] When people have better access to green space, they have better mental health, with lower socioeconomic inequality in mental wellbeing and healthier immune systems, with a reduction of inflammatory based diseases like asthma. The greenest areas have reduced incidence of obesity and type 2 diabetes, while exposure to natural environments has been linked with more favourable heart rate, blood pressure, vitamin D levels, recuperation rates and cortisol levels (lower stress). Quite a hefty set of benefits then.

Unfortunately, whether we have access to green space nearby or not, many of us still spend an enormous amount of time indoors – over 90 per cent of our time on average.[7] How we conceive of ourselves in relation to the natural world determines whether we bring the benefits of engaging with nature into the rest of our lives. Miles Richardson's research quantifies this attitude of connectedness to nature and explores how it can affect our wellbeing. The research involves questionnaires that seek to understand the extent to which a participant's self-identity is independent from or intimately connected to nature. There are a range of different measures with names like 'Nature as Self Index' and 'Connectedness to Nature Scale', although comparisons between the various metrics often find they are closely related to each other, measuring broadly similar aspects of our self-identity. For example, if you respond to questions like 'I often feel a sense of oneness with the natural world around me' or 'I think of the natural world as a community to which I belong' with 'strongly agree' you would likely score high on all the different connectedness-with-nature indices.

These surveys are usually followed with questions aiming to quantify various other aspects of attitude or behaviour. In this way, traits such as autonomy, personal growth, purpose in life and a positive emotional state have all been found to be associated with a person's connectedness with nature, in addition to generally higher self-reported happiness and lower anxiety. If such results remained as statistical associations, we would not be in a robust position to inform the development of these positive traits, but the research field has moved well beyond correlational studies to empirical approaches which can test causality.

The results of new studies raise the intriguing possibility of healing mental illness through interventions that alter people's mindsets regarding their connectedness with nature. Such ideas are on the verge of being widely accepted. Though the biopsychosocial model of health, which links a person's wellness to a combination of biology, psychology and social factors has been around since the 1970s, clinical medicine still labours under a biomedical model, focusing purely on biological factors.[8] Thus, mental illnesses such as depression are treated mainly with drugs (for example, a staggering 13 per cent of Americans over 12 years old and 16 per cent of English adults take antidepressants)[9] without dealing properly with root psychological, environmental and social causes. All that could soon change, and there is a growing awareness of the potential for both 'social' and 'green' prescriptions for health (for example, connecting people to local community services or helping them get active outdoors in a natural environment), although it will need a revolution in healthcare policy to fully integrate these into our national health systems.

It is fascinating that feelings of nature connectedness – of 'oneness' with the natural world – were formerly seen as beyond the remit of scientific studies, falling firmly within the domain of spirituality or religion. Yet, in environmental psychology, we see a coming together of science and spirituality. The split between these two domains stems back to the Cartesian worldview promoted by René Descartes, where mind and the body were seen as being distinct. Such an artificial dichotomy has plagued us in the modern age, not only limiting medicine but also in how we understand spiritual experiences.

Even at the turn of the twenty-first century, science and religion were described as two 'non-overlapping magisteria'.[10] The biologist Stephen J. Gould who proposed this was a natural scientist and perhaps might have had different a view if he was privy to neuroscientific research studying, for example, the brain patterns involved in moral decision making, or the neurobiological correlates of spiritual experience. Suffice to say, many disagreed with Gould's interpretation of the non-overlapping nature of science and religion. The founder of the Human Genome Project Francis Collins wrote that 'in fact, the two magisteria bump right up against each other, interdigitating in wondrously complex ways along their joint border. Many of our deepest questions call upon aspects of both for different parts of a full answer.'

The scientific study of our connectedness with nature is a perfect example of this 'interdigitation' of the two magisteria of science and spirituality. A sense of oneness with nature during moments of deep introspection is not just relevant to the domain of spirituality, as would once be deemed. The feeling of authenticity that such moments bring – as though we have returned home to some deep truth – would previously have been put down as a subjective feeling that we cannot explore objectively. Yet science is now developing ways to understand these states of connectedness. Experiments measure changes in the brainwaves of people such as Tibetan monks as they meditate. The feeling of unity they subjectively report – where the sense of being a discrete entity within the boundaries of our skin marvellously dissolves away – is now supported by science to be the true state of our existence, as the evidence you read in this book will hopefully convince you. We see in the twenty-first century a genuine synergy emerging between science and spirituality. Where there was before a conflict (think Darwin's struggle against the creationists and, in the opposite direction, Gould's rejection of religion as inaccessible to scientific study), we finally see the emergence of what one might call evidence-based spirituality. We can cut through the dogma and fables of organised religion and instead explore scientifically the moral and spiritual states that are often narrowly dubbed 'religious experiences' but may actually be just one possible way to see beneath the veil of self-delusion. An attitude of connectedness with nature with all its attendant benefits for mental and physical health is a more permanent trait-based manifestation of this spontaneous feeling of unity with the natural world.

Perhaps an indication of the growing recognition of the overlapping nature of the two magisteria of science and religion is the statement from the Pope in his 2015 encyclical *Laudato si*. Pope Francis boldly described how Christianity has in the past misinterpreted scriptures to propose that man has dominion over nature. In fact, he emphasises, human life is an integral part of the natural world ('It cannot be emphasised enough how everything is interconnected'), and a new partnership is needed between science and religion in order to deal with the destruction of our natural world.

While this research into our interconnectedness with the natural world develops, another parallel research field is studying our interconnectedness with other people and the advantages this can bring. Let's now focus on a third and final researcher who is another key player synthesising the latest research in a domain of human interconnectedness.

While Miles Richardson watches the evening sun dip behind a cloud gathering the darkness around it like a great coat, at the same moment, on the other side of the Earth, the sunrise races across the Pacific Ocean at around 1,000km per hour and eventually reaches San Diego, California, painting the city with long slants of fresh dawn light. At San Diego State University, Professor Jean Twenge looks forward to the start of her day. Twenge is a prolific writer of over a hundred scientific publications and several books, often focusing on the worsening state of mental health of modern teenagers. For several decades she has produced and summarised new research on how the psychology of humans is changing over time.

A key finding in this research is that levels of individualism – perceptions of the self as self-directed, autonomous and separate from others – have increased over recent generations all around the world. An impressive study led by three Canadian researchers measured changes in individualism across seventy-eight countries since 1960.[11] They considered people's individualistic values through survey responses (about the importance of friends versus family, preference for self-expression and views on the importance of children being independent) and also on practices in each country associated with individualism, such as whether people tend to live alone and the frequency of divorce. The researchers found that for both individualistic values and practices, most countries showed significant increases over time. Even so, there remained evidence of cultural differences between countries, with a few countries declining in individualistic values (Armenia, China, Croatia, Ukraine and Uruguay). Over three-quarters of countries, however, have shown a steep rise in individualistic values and practices.

By itself this might not be a problem; in concert with these changes many countries have experienced higher levels of socioeconomic development. The problem comes though at the extremes, when a healthy self-esteem becomes distorted into narcissism – a collection of traits such as arrogance, conceit, vanity, grandiosity and self-centredness that are damaging to others and the environment. From her base in San Diego, Twenge has studied narcissism using a survey called the Narcissism Personality Inventory (NPI), which was first tested on American college students. Since its inception in the 1980s the test has been repeated many times, allowing Twenge and colleagues to assess personality changes in young people over time. Their worrying finding is a dramatic rise in narcissism continuing up to the present day.[12] It should be emphasised that the NPI doesn't directly measure the most severe, clinically recognised form of narcissism – Narcissistic Personality Disorder. This is a psychological condition from early adulthood in which people have an inflated sense of their own importance, an excessive need for attention and admiration, and a lack of empathy for others. A subset of people with high NPI scores suffer from this severe clinical disorder and, in both cases, occurrence tends to be prevalent in younger people and has grown much more common over time.[13]

Narcissism is problematic; it generally leads to troubled or harmful relationships with others, especially those who are close to the narcissist such as family members or romantic partners. A lack of empathy means narcissists will use others for their own gain with no concern about their impacts on them. Growing trends in individualism and narcissism also correlate with less care about the wider environment. Analysing huge data sets of American new college entrants and seniors, totalling over 9 million respondents, Twenge and colleagues found that civic orientation (interest in social problems, political participation, trust in government, taking action to help the environment) has declined since the mid-1960s, with some of the largest declines in taking action to help the environment between those in Generation X (born 1962–1981) and Millennials (born after 1982). Some of the results of this study in changes in life goals of students across generations are staggering. For example, from 8.7 million respondents, 45 per cent of the Baby Boomers (born 1946–61) had a life goal of being well off financially while 73 per cent wanted to develop a meaningful philosophy in life. By the Millennial Generation these life goals had flipped to 75 per cent of students wanting to be well off financially and 45 per cent wanting a meaningful philosophy in life.[14] It makes you wonder why you would want money over a sound philosophy for happiness. It seems to be a case where the proxy for happiness (money) is confused for the true target. A similar shift at the macro-scale seems to have occurred in our nation states – most countries (except a rare few like Bhutan)[15] have a blind focus on maximising GDP as a proxy for wellbeing of citizens and ignore more direct measures. If GDP is highly correlated with wellbeing that generally works OK, but in some cases growth in GDP comes at the detriment of wellbeing (for example when it leads to deterioration of the environment and impacts people's health).

Perhaps explaining this focus on wealth accumulation and loss of civic orientation, narcissists feel it is acceptable to cheat to get ahead in the world (with cheating in high school and college exams on the increase) and they feel less worried about polluting the environment or taking more than their share of resources, leaving less for future generations. A resource game designed by Keith Campbell, one of Twenge's long-term collaborators, asked participants to roleplay as representatives of a forest company tasked with harvesting timber from a forest.[16] Participants with high scores for narcissism harvested timber less sustainably in initial rounds, exceeding the forest renewal rate and resulting in destruction of the forest and less timber harvested overall. Campbell concluded that narcissism provides short-term benefits to the self but long-term costs to others. The 'Tragedy of the Commons', where humans struggle to balance public versus private gains and fail to protect shared public goods, is amplified greatly when the people are narcissists.

The global spread of narcissism around the world is likened to a disease epidemic by Twenge. But why is it happening? There are many possible causes: the way we raise our children both at home and school, with excessive focus on boosting self-esteem, the way we are surrounded by a narcissistic culture in the media[17] – the cult of celebrities and the glorification of money, power and fame at the expense of more meaningful yet humble community-focused pursuits. When this culture permeates us, it is hard not to slowly absorb it. If someone tells us we need to work on our 'personal brand', we may have once rolled our eyes, but nowadays we don't blink at the use of such language. Just take a step back for a second, though, and think about how odd it is to try to create a brand of 'yourself' – an entity which is constantly in flux and inextricably connected to everything else in the universe. It is simply impossible, and any success you do have creating some kind of image that sticks in other people's minds will not be a true reflection of you, therefore creating an unhealthy (and mentally stressful) dissonance between yourself and the 'brand'.

So what can we do about this? Psychological research is now much more keyed in to understanding how an interconnection with others is healthy and how it can be better encouraged. When the process of interconnectedness works well it generates the important feeling of empathy, where we allow our self-identity to soften and imagine what it is like to experience the world in someone else's shoes. Researchers find that narcissists are unable to express empathy, their self-boundaries are brittle and they cannot see beyond the illusion of themselves as distinct (and superior) to others.

Such understanding provides insights for how we might reverse this pathological disorder, or at least prevent it from first developing in the minds of young adults. Neuroplasticity allows us to change the structure of even mature brains with training, just like we might train physical abilities such as playing tennis or swimming (which after all rely on altered neural networks as well as muscle development). Scientists and mental health practitioners are developing new ways to train minds so that they include others as part of the self-construct leading to increased empathy. Perspective taking is one such approach, where people learn to perceive a situation or understand a concept from an alternative point of view, and it seems possible to train the varied the mental functions that underpin it. For example, researchers have shown how cleverly designed computer games that reward perspective taking can increase empathy. One game involves a space-exploring robot crashing on a distant planet. In order to gather the pieces of its damaged spaceship the robot needs to build emotional rapport with the local alien inhabitants who speak a different language but have human-like facial expressions. Children who play the game for several weeks show greater connectivity in brain networks related to empathy and perspective taking, and increased development in neural networks commonly linked to emotion regulation.[18]

It is not yet known whether such games would work for adults whose neural circuits and sense of selfhood are more strongly fixed, but other techniques may also have potential in breaking down the boundary with others. The UK mentalist and illusionist Derren Brown demonstrates one such approach in his recent TV show, *The Sacrifice*. The participant is an American man, Phil, who has very strong views on Mexican immigrants, verging on racism. After seeing the results from a DNA test, showing that he has ancestors from all over the world, Phil is asked to sit in a room with a Hispanic man on a chair opposite him. For four minutes Phil must stare into the stranger's eyes without speaking. By the end, he is in tears and asks the stranger for a hug. He finds the task forces him to reconcile his abstract views of the other person and, through the intense direct contact, breaks into a state of empathy.[19]

The empathy induced by these new approaches can also lead to real positive changes in the world – other studies show that as empathic concern for others grows, so does the frequency of prosocial (helping) behaviours. So it appears there may well be an antidote to narcissism, but it requires us to better understand the science behind our interconnectedness to others.

This theme has run through the triad of case studies presented here: state-of-the-art scientific advances in understanding our human interconnectedness – whether it be with the microbiome within us, the natural world around us, or with other people – open up considerable benefits in terms of personal and planetary health.

16

A confluence of connectedness

Besides studying the benefits of understanding our interconnectedness across different dimensions, what else do the three researchers in the previous chapter – Proal, Richardson and Twenge – have in common? The DNA code that organises their bodies is over 99 per cent identical between the three of them, and the structure of their brains is similar due to a shared evolutionary history and common aspects of the cultures they grew up in. Beyond this, the actions of these three people connect them together too. They are key players in the great distributed network that represents the sum of human knowledge. Each of these three researchers works at the frontier of an academic field drawing on new knowledge about our human interconnectedness, and they all share a skill in synthesising detailed information and translating it to others who are less familiar with their specific area of study. People like this are key to advancing human knowledge because they act like 'super-nodes' linking different parts of the vast library that represents the sum of human knowledge stretching back over history.

So the three researchers share an awful lot in common yet, intriguingly, there are also further hidden links between them that remain unexplored. The science of our human interconnectedness is still fragmented and there may be undiscovered prizes waiting in its synthesis: first, it is becoming clear that our inner connectedness affects our capacity for outward connectedness – our microbiome affects our psychological self-constructs, so there are *interactions* between our inner and outer ecosystems. This is a new area of research, although the physical structures underpinning these links have been known about for some time. Early human physicians knew that our guts are closely linked to our nervous system by a nerve cord which extends down our spine from our brains and spreads into fine filaments around our abdomen (the vagus nerve). There is a direct and immediate electrical highway by which our microbiome affects our brain functioning. Further gut–brain links result from the microorganisms in our gut producing neuromodulatory chemicals that enter the bloodstream and induce systemic inflammation responses in our bodies. Around one third of small molecules in our blood are thought to originate from the human gut microbiome – right at this moment, there are considerable numbers of bacteria, and the chemicals they produce, coursing through your veins.

All this means there is the clear potential for our gut microbiome to affect our thoughts and emotions. Let's imagine a patient – we'll call him James – who eats a poor diet, full of heavily processed foods and who has a disrupted gut microbiome from eating at odd times of day, often very late in the evenings. This diet, accompanied by stress at work and a lack of exercise, has left him vulnerable to infection that he has treated using antibiotics, many prescribed to him unnecessarily because the infections were actually caused by viruses, towards which antibiotics are ineffective. James now suffers from dysbiosis of his gut microbiome – an unbalanced composition of microorganisms – and this is likely to be partly responsible for his feelings of fatigue and anxiety.

Researchers have known for some time how bodily sensations provide the critical basis for our emotional experience. It is thought that the complex web of neurons in the gut, linked to the brain via the vagus nerve, can act as a memory store for intuitive emotional states (giving some real meaning to the phrase 'I have a gut feeling about this').[1] These feelings may also have a role in mediating social interactions, including feelings of empathy. Our interoceptive awareness – the awareness of our bodily sensations – also affects how we feel and the decisions we make.[2] In James's case, it could even result in his withdrawing from social company, because it turns out there's a close relationship between negative feelings in our gut (including nausea and abdominal discomfort) and social withdrawal.[3] Experiments have demonstrated clearly how gut status can mediate emotions – for example influencing how the brain deals with sad emotions when listening to a piece of music. To understand how the status of our guts influences our emotions one experiment involved intragastric infusion (putting food directly into the gut without making the participant eat it which could affect the emotional response).[4] Participants' stomachs were filled up with either fatty acids or a saline solution (as a control), and they were then played certain classical music tracks known to stimulate sad emotions. It turns out that having fatty food in our guts reduces the impact of sad emotions. Voilà, the science behind comfort eating!

So James's depression might be caused in part by the disruption of his microbiome composition. This has led scientists to suggest certain foods which encourage the growth of beneficial bacteria in the gut and restore a healthier community balance may help in some mental health problems. One day soon these psychobiotic foodstuffs, tailored to our individual microbiome and in combination with an improved dietary and exercise regime, may be able to replace standard antidepressant drugs.[5]

Is it possible that the state of our microbiome, by mediating gut-related feelings, could also affect empathy and narcissism, linking the research spheres of Proal, Richardson and Twenge? The idea is not so far-fetched. Autism-related disorders are often associated with both a low capacity for empathy and a disrupted microbiome, opening up nutritional approaches to treating them.[6] Furthermore, our internal body feelings, in particular those related to our guts, determine our capacity for feeling rooted and relaxed and achieving a less 'object-centred' state of attention, allowing the boundaries of our self-identity to become expanded – a likely precursor to achieving greater social awareness and empathy.[7]

There are also links between our brain activity and microbiome that work in the opposite direction: when our minds become stressed, the hormone cortisol is produced by the adrenal gland in our brains and streams through our bodies. This affects immune cell activity and gut permeability that, in turn, influence the composition of our microbiome. Experiments on animals such as rats and monkeys, where offspring are stressed due to temporary maternal separation, find disruptions in the gut microbiome of these animals.[8]

Our attitudes and behaviours affect our microbiome in other ways too: through our dietary choices and where we choose to spend our time. People who spend more time in natural environments (as those who feel connected with nature do), are exposed to a greater range of microorganisms and have a more diverse microbiome as a consequence. In contrast, for many who lead highly urbanised lifestyles, we have lost exposure to microbial diversity and this seems to be reducing our microbiome diversity. To give just one example, allergies such as hayfever are rarer among farmers in many countries, as well them having reduced incidences of other autoimmune diseases such as inflammatory bowel disease. There is now a global initiative, The Healthy Urban Microbiome Initiative, aiming to improve the design of cities to ensure exposure of urban populations to healthy environmental microbiomes.[9] So there is a two-way feedback at play here – on the one hand how we think and act affects our microbiome, and on the other, how the state of our microbiome alters our thoughts and emotions and our ability to feel empathy.

If the human microbiome can affect empathy – our ability to connect to others – it probably also affects our connectedness with the natural world around us, because it turns out that the two psychological constructs share similar aspects. Early environmental psychology research treated self-identity with others and with nature as separate domains. Values related to self-identity had been labelled as either *egoic* (a focus on an isolated 'me'), *social-altruistic* (valuing other people), or *biospheric* (placing value on plants and animals). There was a rich seam of research into how close relationships with other people, and the ability to take on another's point of view related to a greater overlapping of self-identity with the other person.[10] Researchers proposed that closer relationships involve a blurring of the boundary of the self with the person (or in technical language: 'an overlapping of self-schemas and schemas of another person'). They developed a scale to measure this – the Inclusion of Other in Self scale.[11]

In parallel, research on how our true self is enmeshed with the fabric of nature was progressing relatively independently in the so-called 'deep-ecology' and ecopsychology literatures. This dichotomy, however, was purely a result of academic fields working along quite independent tracks.[12] Later, it became clear there was little support for the distinction between the *social-altruistic* and *biospheric* clusters of self-identity. Instead, according to environmental psychologist Wesley Schultz, the two cannot be differentiated from a more generalised 'self-transcendent' cluster that reflects the degree to which a person includes other people *and* other living things within their notion of self.[13] This suggests a potential bridge between research on social empathy and connectedness with nature. If Miles Richardson and Jean Twenge were to chat on the phone about their research, they would likely find a lot in common.

This parallel investigation of different research fields into what might essentially be related aspects of psychology explains why metrics the two fields have developed turn out to be closely associated with each other. For example, there are positive associations between people's connectedness-with-nature scores and those on tests designed to assess empathy. So fostering a sense of connection with nature should also address the declines in empathy we see in our individualistic culture. There are also links between connectedness with nature and narcissism, especially in relation to narcissistic traits of exploitativeness and entitlement.[14] Narcissism is a major barrier to solving environmental problems, so developing new ways to help people connect psychologically to nature may offer a way to solve the environmental problems and concurrently reduce the interpersonal cruelty that arises from the growing narcissism 'epidemic'.

In the early twenty-first century, we stand on the precipice of a new integration of these different aspects of human interconnectedness – a confluence of connectedness research. Through new research developments into the human microbiome (connectedness with our inner ecosystem), environmental psychology (connectedness with nature) and in studies of empathy and narcissism (connectedness with other people), we are learning how to better manage ourselves and the natural world around us. New scientific advances at the intersections of these research fields offer the beguiling promise of convergence towards a more unified understanding of our interconnectedness with the world around us. This will deliver significant personal and planetary benefits at a time when we urgently need them.

17

How a network identity leads to global responsibility

Every nation and every man instantly surround themselves with a material apparatus which exactly corresponds to their moral state, or their state of thought. Observe how every truth and every error, each a thought of some man's mind, clothes itself with societies, houses, cities, language, ceremonies, newspapers. Observe the ideas of the present day ... see how timber, brick, lime, and stone have flown into convenient shape, obedient to the master idea reigning in the minds of many persons   Ralph Waldo Emerson

Patricia Smith presses the lever switch firmly downwards and the machine buzzes. An electric hiss in the room opposite is shortly followed by a long, drawn-out human scream. 'Please, please stop,' shouts a voice, 'I can't stand the pain. I have a heart condition!' Patricia squirms guiltily in her chair, looking up at the scientist in a lab coat standing beside her. She sighs with deep resignation and turns back to the list of paired words, from which she must read one of each pair and the person in the opposite room must recall its partner. Patricia continues to

read from the list, but almost instantly the experimental subject has made another mistake. She looks at the lever in front of
her. The dial is labelled '375 VOLTS (DANGER: SEVERE
SHOCK)'. 'I can't ...' she begins to say, but the scientist in-
terrupts her. 'Please continue. The experiment requires you to
continue.' Patricia rubs her hair repeatedly and puts her hand
to her mouth – clear signs of stress. She looks at the scientist
again then meekly presses the lever button. A loud cry emerges
from the next room but is suddenly cut short to give way to an
eerie, guilty silence. 'What have I done?' Patricia says brokenly.

The above events (but not the participant's name) are based
on an actual experiment designed in the 1960s by American
social psychologist Stanley Milgram. In the experiment, participants – just normal people like you and me – pressed a
switch to administer what they believed were electric shocks
up to lethal voltage levels to unseen subjects. It was not just
one or two people who proceeded to knowingly increase the
voltage to harmful levels – 65 per cent of participants, under
the stern urging of the attendant scientist, continued to administer shocks up to the highest level. A repeat of the study in
2016, using state-of-the-art techniques to reveal what is going
on inside participants' brains at the time of the experiment,
showed *lower* brainwave activity in crucial areas for autonomous decision making.[1] So rather than worrying about the
moral transgressions of their choices to pull the lever, it seems
that the participants were *deferring* the choice to the scientist
advising them.

These experiments are a critical contribution to science
because they begin to explain how humans in extreme

circumstances, such as soldiers in times of war, can be ordered to conduct acts of atrocity even though it goes against every sinew of their moral fibres. This research may also explain many minor everyday moral transgressions.

How often do we defer moral judgements that we ought to make ourselves? We probably do this pretty much every time we spend our money to buy something. Take the example of battery-caged chickens. If we had to rear them in our own gardens, not many people would be prepared to subject birds to the inhumane conditions they suffer in battery farms, where they have to survive in an area less than an A4 sheet per hen, with little or no sunlight, and where beaks are cut to prevent cannibalism and self-harm. Yet, many of us still buy products that contain eggs and meat from such chickens. For example, in 2014, 95 per cent of US eggs were produced in battery cages.[2] Our purchases are at one end of a causal chain, the other end of which are harsh acts of animal cruelty. Many of our purchases also lead to violence against the wider environment that we would never countenance carrying out directly. Hundreds of household products contain palm oil, whose production involves clearing and total devastation of millions of kilometres of primary rainforest. If it were us sat in the bulldozer driving seats, we would directly observe the terrible destruction of homes of countless animal and plant species, and might question whether this was worth the slightly cheaper bar of soap we bought.

Let us return for a moment to the idea of causal chains. We mentioned acts of atrocity committed under orders during wartime, a clear example of which would be German soldiers ordered to murder millions of Jews by sending them into gas chambers during the Second World War. Following some of the scientific research started by the Stanley Milgram experiment and continued by neuroscientists and psychologists since, we can at least better understand how the soldiers ended up taking these actions, even if we cannot necessarily absolve them. Because the parts of their brains making autonomous decisions were subdued (and one might argue that much of the training to become a soldier is specifically so this occurs), it is the superiors of these soldiers who gave the orders who are ultimately responsible. Therefore, the heads of the German SS should receive correspondingly more severe punishments in war crime tribunals. Now, let us return to the less sinister, but more widespread example of battery chicken farming. When we purchase our bargain basement-priced chicken products, we order from a supermarket who then orders a farmer to produce eggs and meat by the cheapest available method. So, should we blame the farmer for the maltreatment, or those further up the causal chain – the supermarket and finally ourselves? In line with the discussion above, the farmer might argue that we made the orders and are ultimately responsible for the consequences of millions of chickens living in inhumane conditions.

Maybe the answer is that we are all to blame and we all share some responsibility. The nineteenth-century French economist Frédéric Bastiat thought so at least. His stance was that: 'When a man is impressed by the effect that is seen and has not yet learned to discern the effects that are not seen, he indulges in deplorable habits, not only through inclination, but deliberately.' The problem of course is that the longer and more complex the supply chain of a product is, the more diffuse is the moral responsibility, and it is correspondingly easier to disown and forget (although it is never negligible). This is a growing problem as our economies become increasingly globalised. Hugely complex supply chains mean that one purchase sends off thousands of ripples across the globe, so fine they are almost untraceable.[3] As the global population becomes urbanised (and we already passed the threshold where most of us are city-dwellers in 2014),[4] we become further disconnected from the raw products that we consume. Former UK government adviser and author Steve Hilton says:

> We live in an age when the effects of our decisions seem less and less important because we don't really know what they are. We 'love a bargain' but we don't see the appalling conditions endured by the people who produce a product that can be sold so cheaply. We troop to the supermarket but don't see the small businesses and farmers whose livelihoods are wrecked by 'everyday low prices'. The human consequences of our actions are felt by people separated from us by time, space and class.[5]

We fail to see these connections because they are not apparent, and we are so wrapped up in our individual selves and small social spheres that we fail to look for them. There is a condition called apophenia where people see patterns where they do not exist.[6] Yet, what about the opposite, of failing to perceive connections that really do exist? There is no term for this systemic blindness (although it seems to be one characteristic of the individuation pathology described earlier). If we suffer from such blindness, is this a good excuse for reneging on the responsibility regarding our consumer choices? Perhaps, but maybe blindness is the wrong term here because it implies some inevitability and permanence of the condition. In most cases, it is akin to having our eyes shut to the implications of what we buy – a wilful ignorance. We don't want to see or hear about the damage because that places a huge burden of moral responsibility on our shoulders. Better just that no one talks about it. In fact, there seems little incentive to expend the effort in finding out about the impacts of our choices, because we know it will be bad news. Better to close our eyes and defer the moral responsibility in the great merry-go-round of global trade networks.

The philosopher Julian Baggini suggests that we do not escape responsibility in this way: 'If I contract a builder who uses slaves, I am just as much at wrong as if I owned the slaves, just as I am if I kill them personally.' In the context of animal welfare, the American essayist Ralph Waldo Emerson asserts: 'You have just dined, and however scrupulously the slaughter-house is concealed in the graceful distance of miles, there is complicity', but the field of environmental ethics is generally surprisingly poorly developed on this topic of responsibility through complex causal chains. Researchers describe two types of violence against the environment: 'agential violence' that is a direct, intentional act of destruction, and 'structural violence' that is indirect and unintentional. The long distance environmental and animal welfare impacts we discuss are often placed in this latter category of structural violence, but isn't 'structural' violence a bit of a misnomer, because it allows us to blame some kind of faceless 'system' for the environmental ills of the world? The 'system' is simply the set of institutions that reflect our combined worldviews (past and present). In reality, environmental violence is the consequence of millions of choices, so we might better call it something like 'diffuse agential violence', which more accurately reflects our eschewed moral responsibilities. One could argue about the intentionality here: as the Milgram experiment and its successors show, we don't engage in the same moral reasoning in indirect interactions as when we are in direct interactions; instead, we defer our moral responsibilities to intermediaries in the chain and confer them a moral authority which is unwarranted. There are also questions about the extent to which violence is an emergent property of a system rather than simply the sum of individual choices. For example, if a driverless car kills a pedestrian who is the person liable: the programmer, the engineer, the road planner, the car retailer? Similarly, with regards to the environmental impacts of a multinational company that causes environmental devastation; who is responsible: the shareholders, the company managers, the employees on the ground, or the consumers? In these cases, one might suggest it really is the structural system that is to blame and there is a lack of intentional agential responsibility here. Yet, the system is comprised of the thousands of small choices of individual people, even if some of them are people that lived in the past and influenced the current set of principles and protocols we now follow. Although excruciatingly difficult, disentangling responsibility in this way to lay a small part at the feet of each agent may be the only way to ensure that such systems act in an ethical way.

We may feel like our choices have minimal effect ('what difference is little old me going to make?'), but the collective impact of our personal choices can be huge. A recent report identified collective human impacts as a major global risk in coming years: our combined diet choices determine health epidemics, multiple small local dams affect regional water cycles, movement of people to cities causes overcrowding and food security issues. Our collective personal consumption choices are ultimately what drive environmental degradation, climate change, poor animal welfare, and a whole range of other issues. The nature writer Michael McCarthy puts it bluntly: 'Nature is not being destroyed by a particular political or economic creed; it is being destroyed by the runaway scale of the human enterprise.'[7] Each of our individual purchases is like a vote for how we treat other people and the natural world, and even if we don't directly see the results, our votes do count, as the well-documented decline of the global environment attests. All our small choices ripple across the surface of the globe and accumulate to create unstoppable tidal waves of destruction. McCarthy once again puts it aptly: 'Most ordinary individuals do not care ... the trouble to come is their own individual choices, multiplied 7 billion times.'

How can we stop this systemic destruction? Many people look to governments. The social contract between the people and their government should be such that governments intervene to ensure that public goods are protected, yet around the world governments are failing in this task. Many politicians are swayed by the lobbying of big multinationals, and the short-termism of election cycles. This, combined with the influence of a powerful business-driven right wing media cabal that criticises tough environmental regulation, means governments are reticent to put in place unpopular regulations that could protect the public in the long term. As a consequence, many argue the social contract between governments and citizens is now broken. Increasingly, governments adopt laissez-faire free market approaches, assuming long-term benefits sought by the public will be reflected in their collective consumer actions (which they clearly aren't), while individuals think there is no need to curb their own behaviour because governments have an eye on the bigger picture and will intervene to protect society's long-term benefits (which they often won't). There is a simultaneous abdication of responsibility, and while both the public and our governments bury their head in the sand and ignore the problem, the global environment slowly withers from the accumulation of billions of small cuts.

It is a very thorny problem because the moral compass of humans hasn't evolved to intuitively respond to harmful impacts on such global and long timescales. While in the last few hundred years our transport and trade networks have expanded to encompass the entire earth, our sense of moral responsibility hasn't kept pace. Much of our morality, whether it is genetic, culturally based, or likely both, evolved when we lived in small groups of families as hunter-gatherers. As we have learnt in earlier chapters, we evolved to respond with empathy to seeing someone close to us in pain; in many cases physiological 'mirroring' mechanisms allowing us to actually feel their pain, prompting us to help them. We then created formalised group rules, and ultimately laws, that prevent us from physically harming others. Yet, in a globalised world where we are connected not to tens, or hundreds, but to billions of unknown people, these mechanisms fail us. Our behaviours in a globalised world become maladaptive. Our moral compasses are not sensitive enough to respond to impacts halfway across the globe that we cannot see directly; or, to put it another way, we put our moral compasses into sleep mode and defer responsibility to others. These long-distance impacts do not trigger the empathetic emotional response that would occur if they were localised. Similarly, our rules and laws are unfit for dealing with cases where violation is caused not by one, or even by several people, but by millions of people, with the web of causality stretching outwards in long tortuous paths.[8] Both our governance and our personal morality are not up to the task at hand when it comes to solving globalised problems such as deforestation, climate change, social depravation and poor animal welfare. So perhaps both need fixing – how could we achieve this?

First, let's look to our sense of morality and consider a revolutionary correction suggested by Norwegian ecologist Arne Naess, who suggested that to solve environmental problems, our conception of self-identity needs to expand from an 'egoic self' to an 'ecological self' that encompasses all of the earth's living systems. While one is working only within a narrow concept of the self, he argues, environmentally responsible behaviour always relies on altruism, which is too inconsistent to reverse the wide-scale environmental degradation driven by the collective human endeavour. Instead, enlargement of self-identity to an 'ecological self', integrating all those organisms we are impacting, can result in environmental behaviour as a form of *self-interest* – care for the natural world beyond our immediate bodies becomes an act of self-love. In Naess's own words:

> Having an extended sense of identification leads us to say that we defend our homeplace as part of ourselves ... We care for our place and others, we come to identify with their need and well-being, and we have a greatly enhanced and larger sense of community and interdependence. Our well-being and that of our community are closely aligned. Thus, we naturally and spontaneously care for our place and seek to protect it. For this we do not need a moral axiology, a set of rules and enforcements held over us to force us to act.[9]

Naess's theory does seem to have something going for it, but it will require overcoming a resilient perspective of the world that has evolved over thousands of years. In evolutionary terms, altruistic behaviour (the act of helping others to our own cost), has generally only evolved in a limited range of circumstances; for example, helping family members who share similar genes (so-called kin selection), or helping others in our close social 'in-groups' from whom we are likely to receive reciprocal benefits.[10] These benefits allowed small human tribes to function coherently and successfully together, but due to the requirements of high genetic similarity or strong likelihood of reciprocity, it is unlikely they could allow for the significant altruism in large groups, especially at global scales needed for global sustainability. Yet, what if we could some-how significantly reprogram and extend our conception of the in-group that we belong to, then perhaps such large-scale cooperation might just be possible?

In psychology, the study of collective group identity has a long history. Many studies have demonstrated how members of groups act more favourably to others in the group versus those outside – they show 'in-group bias'. Such favouritism, however, has downsides and is the cause of many callous acts of violence throughout history (national wars, racial conflict, football hooliganism). Individuals are capable of horrific cruelty to those outside their chosen 'tribe'. It is not yet clear whether Naess' suggestion to conceive of humanity and life on earth within a single tribe could operate to harness the same power of collective identity when there is no outgroup as a counterfactual to stimulate in-group cohesiveness – can you have 'in-group' benefits without an 'out-group'? However, there is some intriguing evidence that Naess may be right.

Despite Naess's pioneering 'deep ecology' writings mostly occurring in the 1970s and '80s, it has taken until the turn of the twenty-first century before research into human-nature connections has really exploded, with a growing number of international conferences exploring the topic. In 1990 a con-ference at the Harvard-based Centre for Psychology and Social Change entitled *Psychology as if the Whole World Mattered* con-cluded with a statement that 'if the self is expanded to include the natural world, behaviour leading to destruction of this world will be experienced as self-destruction'.[11] From initial philo-sophical theorising, research is rapidly moving towards a much more objective empirical approach, with whole new branches of environmental and conservation psychology springing up. As described earlier, researchers in these new fields empirically quantify people's connectedness to nature and relate this to the frequency of pro-environmental behaviours. A recent review of the science explores how such human 'connectedness with nature' can affect environmental outcomes through influencing people's behaviours and actions.[12] Matthew Zylstra and the other authors of this review define connectedness with nature as a state of consciousness 'comprising symbiotic cognitive, affective, and experiential traits that reflect, through consistent attitudes and behaviours, a sustained awareness of the interrelatedness between one's self and the rest of nature'. Through the carefully designed questionnaires that allow participants to be given a connectedness-with-nature score, researchers show that people who score highly in this characteristic are also much more likely to be involved in environmentally responsible behaviour (as well as showing other benefits such as reporting themselves to be happier, as we learned previously). Studies have found that pro-environmental behaviours, such as recycling, reducing waste and carbon dioxide emissions all follow from an attitude of greater connectedness with nature. For example, Australian farmers who score highly in terms of their connectedness with nature are more likely to manage their farms in an environmentally friendly way.[13] This makes good sense if you think about it, because when our sense of self-identity is integrated with the natural world then harming that world becomes in effect equivalent to self-harm. Encouraging pro-environmental actions, therefore, does not rely on the difficult challenge of culturing altruism – helping others at the expense of yourself – but by simply encouraging wider extension of a natural attitude of self-care.

It is possible to train and cultivate this perspective of connectedness with nature by encouraging outdoor experiences (for example, outdoor sports and nature recreation, facilitated eco-adventure and field trips, unstructured extended nature immersion). Zylstra and colleagues describe such an approach as 'strategic mentoring as part of a culturally embedded process', which, in plain English, means helping each other to overcome the self delusion that we are discrete, independent entities. Naess would be pleased to see this research, documenting scientifically how an 'ecological self' can be cultivated and how it does indeed lead to a reduction in our impacts on the environment.

The second constraint beyond personal attitudes is of course our governance. We may rue the fact that even in the mature democracies of the world there is little long-term thinking and planning to adequately protect the environment. It is easy to blame the 'system' here, but we ought to remember that the system, comprised of legal, economic and political norms, is really just the outcome of our combined human perspectives organised into a set of collective institutions. The most effective way to change these institutions is to tackle the root of the problem: our collective human perception of the world around us. Anything less will be ineffective. The writer Robert Pirsig in his novel *Zen and the Art of Motorcycle Maintenance* makes exactly this point in response to the rapidly growing counterculture in the 1960s which made naive calls to tackle the 'system', or to deal with the 'man'. Pirsig saw such ideas as pointless, because there is no one steering our great lumbering socio-economic system; its direction is simply the combined outcome of individual values. The 'system' is constituted from billions of people's actions, even if we fail to comprehend the complex web of causality that underlies it. As the protagonist in his novel explains:

> To tear down a factory or to revolt against a government or to avoid repair of a motorcycle because it is a system is to attack effects rather than causes; and as long as the attack is upon effects only, no change is possible. The true system, the real system, is our present construction of systematic thought itself, rationality itself, and if the factory is torn down but the rationality that produced it is left standing, then that rationality will simply produce another factory. If a revolution destroys a systematic government, but the systematic patterns of thought that produced the government are left intact, then those patterns will repeat themselves in the succeeding government.

Hence, if we fail as individuals to appreciate our interdependent relationship with nature, we cannot expect an economic system to take into account such interconnections and protect the natural world.[14] If we fail to take personal responsibility for our cruelty to animals in complex supply chains, we cannot expect our political institutions to develop appropriate regulation against such cruelty; and if we fail to find our true common ground with other people, we cannot expect our justice system to treat criminals with the compassion they need.[15] These are just a few of many examples prompting the need for a fundamental change in our mental models regarding our place in the world – we need a fundamental realignment of self-identity, from seeing ourselves as isolated individuals to being part of a deeply interconnected system. Only then can truly progressive change be precipitated upwards into institutional arrangements that afford genuine sustainability.

There are some success stories to incentivise us. A widening appreciation of the devastating effects of climate change led people in the UK to pressure the government into developing one of the most progressive climate change regulations, the Climate Change Act 2008, that has made it legally binding for the country to reduce greenhouse gas emissions to at least 80 per cent lower than 1990 levels by 2050. This, along with efforts by many other countries, set the scene for the 2015 Paris Agreement on Climate Change. In the face of the climate strikes by school children occurring in many countries across the world in 2019, the possibility of extending this target to 100 per cent emission reductions by 2050 or even earlier is becoming politically possible.

Things can also slip in the wrong direction too though. A recent shift towards right-wing thinking across much of the USA (which tends to shrink the in-group that people associate with, in this case 'America first') led to the Trump administration pulling out of the 2015 agreement on climate change and effectively hamstringing the Environmental Protection Agency. Other countries are not immune to this right-wing shift, towards the end of the same decade, many are increasingly dominated by politics that promote national interests before global responsibilities – for example, refusing to help international migrants, many of whom have been driven by their homes partly by climate change primarily caused by developed nations.

In the face of such clear links between global environmental problems and the mindset of individual people, our perceptions of self-identity are set to become the battleground for environmental and social campaigns of the future. Recent syntheses of global problems, such as that by the World Wildlife Fund in their *Living Planet 2016* report, have begun to take a holistic, systemic approach. They conclude that mental models of the world – reflecting the beliefs, values and assumptions that we personally hold – influence the design of system structures, the guidelines and incentives that govern behaviours and, ultimately, the individual events that make up the flow of daily life.[16] The leverage points that can lead to genuine lasting change, therefore, are at the level of these underlying mental models – we first need to change ourselves to change the world.

It is often said that when we are young and optimistic we strive to change the world around us, but when we are older and wiser we realise the futility of this and aspire to change ourselves instead. *Yet*, to solve the significant environmental problems the world now faces, we actually need to do both – to change the world *and* ourselves. *Yet*, it is even more nuanced than that – it turns out that changing ourselves is a prerequisite for changing the world.

Although environmental problems are caused by both

inappropriate governance and the limited scope of our individual morality, they share a common solution – to develop and act in line with an accurate understanding of our interdependence with the world. Of course, saying is often a lot easier than doing, but there are rays of hope in how changes in our mindsets and belief systems are beginning to be quantified and our scores on such metrics can be improved with training. It may be that we are on the cusp of a transformation in our practical understanding of how to develop such a perspective. Then we may objectively chart our progress towards a new paradigm where people fundamentally acknowledge and take responsibility for their impacts on others and the natural world. This emerging connected consciousness will support a new intelligence which exposes the once-hidden pathways by which we impact the world around us. If the Milgram experiment were repeated in another twenty years, we would see people refusing to defer responsibility for their actions.

18

Your mind is a battleground

> Even the technology that promises to unite us, divides
> us. Each of us is now electronically connected to the
> globe, and yet we feel utterly alone.
>
> Dan Brown, *Angels and Demons*

On a clear day in the Himalayan mountains every view you've
ever seen before seems to have been through the lens of mis-
prescribed glasses. The air is so crisp it feels like you could
reach out and touch the peaks of the mountains that jostle
on the horizon. In the foreground, the grasses climbing their
steep flanks flow in waves as the breeze glides through them.
Tenzin, an old Buddhist monk, sits cross-legged on the floor of
his temple perched high on the slopes. Eyes closed, a contented
smile on his face, he rests in a state of poised equilibrium, his
mind deeply rooted in his body and in a state of emptiness,
so spacious it seems to internalise the vast mountains around
him. When he begins to meditate on compassion, he cultivates
feelings of deep love and empathy for all of humanity, and although he sits in this remote location, isolated from the other
7 billion humans on the planet, psychologically he feels a deep
and meaningful connection to them. Indeed, he feels a strong

sense of kinship and deep responsibility to all the plants and creatures of the natural world.

Tenzin has very little ecological impact on the world. The Buddhist practice of simply noticing but not responding to desires or slavishly following urges, such as those to purchase non-essential goods, seems alien to most of us. The average European consumes over thirteen tonnes of raw material, and energy equivalent to three tonnes of oil, while producing greenhouse gases equivalent to nearly nine tonnes of CO_2 each year.[1] Things weren't always like this. For 99.9 per cent of the time since the evolution of modern humans around 250,000–300,000 years ago, our consumption of goods was minimal, limited to essential food and fuel.[2] Over the last sixty years, dubbed the 'Great Acceleration' by some researchers, near exponential increases in consumption-related variables such as energy and water use and fertiliser production have led to extensive impacts on the natural world. The globalisation of our economies has spun a complex web of causal influences. If we want to live ethically and sustainably, this web needs to be managed responsibly. By simply flexing a few fingers at my computer keyboard I can order two dozen battery hen eggs, perpetuating the life of misery for a caged chicken. At my ultimate leisure, I can command trawling of the Pacific Ocean, destroying precious sea-bed habitats.

As well as global economic connectivity, we now have digital connectivity across the entire earth with our information technology, which brings yet more power and personal responsibility. With seemingly little consequence, I can unthinkingly forward on a meme via social media; yet it is the accumulation of such acts that lead to bullying and suicides, and can even change the course of democracies.[3] If humans living hundreds of years ago could see us now, we would probably seem like networked demi-gods to them, wielding great power of economic and digital connectivity, albeit often carelessly.

Simultaneous with this great power, we have become less psychologically connected with one another and the natural world around us. It is a cruel twist of irony that the advent of information technology in the modern world is coincident with rising levels of mental health disorder, including anxiety depression and self-harm. While Tenzin the Buddhist monk feels psychologically connected when physically secluded, many of us in the modern world are psychologically isolated while digitally and economically connected.

And we can't easily go back. Despite the allure of dispensing with our smartphones and closing our national borders in economic protectionism, such acts are ultimately self-defeating and unlikely to last. If you imagine two axes of a graph – psychological connectivity versus material connectivity (comprising both economic and digital technology), Buddhist monks like Tenzin sit in one quadrant – with high psychological connectivity and low material connectivity – while most twenty-first-century modern humans sit in the opposite quadrant. What about the other two quadrants? One is having no psychological or material connections – a lonely secluded life of solitary confinement. The other is being both very materially and psychologically connected. In this state, we could use those material resources as opportunities to do great good for the benefit of those who we connect and empathise with.

Think of fair-trade networks that enable economic trade while upholding social justice, or the crisis mappers in Chapter 7 who volunteer their skills in state-of-the-art information technology to help complete strangers involved in catastrophes on the other side of the world.

In light of this applied compassion, the inactivity of the kind but secluded monk on the mountainside might seem like a wasted opportunity (though even the most active humanitarian is likely to need at least some solitude to recharge their batteries). An important question then is how do we enable the transformation of humanity to the quadrant where they remain networked to each other while also becoming more emotionally connected and caring? In previous chapters, we learnt how having a sense of self-identity which overlaps with others and with the natural world is key to allowing the pro-social and pro-environmental actions that naturally follow. So how do we transform our mindsets in this way?

Many have sought methods to transform their consciousness through the ages. As we have encountered, some routes offer the allure of shortcuts – intoxication using psychoactive drugs, or hallucination through starving or overloading the brain with oxygen. When research on LSD was made illegal, the Czech psychiatrist Stanislov Grof developed a technique he called Holotropic Breathwork involving rapid breathing to enter a supposed state of transpersonal consciousness, but there were concerns the technique could cause seizures or lead to psychosis among vulnerable people. Even if such shortcuts were safe and effective in leading to a widening of the sense of self, they are generally only temporary. The mind is like a muscle from which we can temporarily coax enhanced performance with mind-altering drugs, but to make significant and permanent changes it needs the creation of long-term structural change. There is only one sure route to do this – repeated practice.

Meditation is one method that is gaining increasing scientific support in terms of its ability to permanently alter the structure of our neural networks. We have touched on this briefly already, so let us now consider the benefits in detail. Meditation leads, for many, to a wide range of positive benefits, including attitudes of compassion and empathy. Measurable physical changes in the brain associated with reduced self-centredness occur after as little as two months, and significant changes continue to accumulate over prolonged practice. Meditation practice decreases the activation of self-relevant regions of the brain when it enters its 'default mode'. This default mode is a region of the brain that becomes active when we are not engaged in some complex task. Rather than resting, researchers have found that our brain often remains highly active but in relation to self-focus and rumination. After just a few days of practising mindfulness meditation, however, the postcingulate cortex region of the default zone – a primary network for self-focused thought – becomes more connected to the dorsolateral prefrontal area of the brain, which is responsible for managing and down-regulating the default brain mode. Further long-term mindfulness practice can lead to an overall decrease in activity in the areas of the default mode related to self-focus. Hence, the mechanism behind our reduced self-obsession through meditation practice is becoming slowly revealed.[4]

Using live brain-scanning techniques, neuroscientists have found the neural pathways related to attention, compassion and empathy, reaction to stress and sense of self, all become restructured in the brains of experienced meditators. Such neural rewiring has impacts not only in terms of beneficial health outcomes for practitioners (reducing inflammatory responses and lowering levels of the stress hormone cortisol) but also in terms of positive interactions with others. Researchers have used clever experiments to explore the effectiveness of meditation techniques devised to promote empathy towards others.

The common problem with studies assessing behavioural changes is that telling participants they are undergoing some process that might alter the frequency of a behaviour primes them to expect exactly this and can bias the results. To overcome this, studies like one conducted by Yoona Kang and colleagues in 2014 use a neat approach that measures how quickly participants pair different types of words.[5] This kind of test is regularly used to assess implicit (subconscious) bias. Kang found that participants who had undergone a course in loving-kindness meditation had reduced intergroup bias. In particular, they were less likely to stigmatise homeless people compared with participants who were simply taught about the value of the meditation technique but not shown how to perform it.

Meditation, therefore, seems an effective and controlled way to realign our self-identity and overcome the persistent self delusion of independence, although it is by no means the only approach. Rather than sitting on our own, we might find that working with others – social learning – can help. Some methods might be familiar to us already, such as green gyms where people work together outside and engage in nature, others might be newer, such as education programmes utilising recent psychological research to grow empathy among school pupils.[6] Others may seem downright zany, for example the Council of All Beings, a ritual where participants dress as plants or animals and from a species-eye-view explain what it feels like to suffer destruction at human hands.[7] Outlandish as it may seem, this kind of perspective-taking has been shown to be effective in changing mindsets and genuinely increases environmental concern. When participants are shown images of animals being harmed, such as seals trapped in a net, and then asked to imagine how the animal felt, they scored higher on tests of wider environmental concern compared with those asked to only analyse the images objectively. The feelings of empathy induced by this perspective-taking increase the overlap between our self-identity and nature. However, I would personally prefer to find a way to obtain such benefits without needing to dress up as a mushroom.

In addition to re-embracing age-old approaches, there is certainly a role for new technologies to facilitate psychological transformation. I have already described the computer games designed (and tested) to improve empathy. Neurofeedback technologies may also help to improve the efficiency of meditation practice by providing live feedback on changes to brain states. In this way, meditators can more quickly learn the most effective way to achieve altered states of consciousness, fast-tracking to the benefits otherwise obtained through long-term practice. Such techniques (and the technology they require) are still in their infancy, but they offer promise in changing our mindsets without the need for thousands of hours of practice. Think how new coaching approaches help modern athletes attain mastery of a sport much quicker and to a higher level than ever before – is there any reason why the same may not be achieved for brain training?

New approaches from psychological research into breaking habits may also help; after all, self-rumination is really no different from bad habits such as smoking in that certain cues lead us to regress into patterns of bad behaviour. If being on social media leads us to greater self-centredness[8] then we can think of proactive tactics to minimise our time on these platforms. Treating our self-identity like a habit means techniques that help reduce the frequency of bad habits can be effective in preventing us drifting into unhelpful mental states. For example, 'implementation intentions' are rules that people plan out ahead for how they will act in a situation when they normally enact bad behaviours ('Whenever situation x arises, I will initiate response y which will help meet my long term goals'). Gradually, the decision-making process to avoid these slips becomes automatised as a good habit. First developed by psychologist Peter Gollwitzer, the approach has been shown to have success in reducing alcoholism, reducing speeding in vehicles and improving the use of green travel options.[9]

In terms of planning ahead to prevent circumstances that foment an isolated sense of self, we may want to avoid certain physiological states. Some food types, very fatty foods for example, are long recognised by certain religions to be less conducive to maintaining the balanced attention required to achieve a self-transcendent state of mind. We can easily recognise this ourselves if we try to carry out tasks requiring deep concentration after eating fast food. Diets affect our capacity in the longer term too. Recent science shows how our diet alters our gut microbiome composition leading to lasting changes in interoceptive awareness, influencing our ability to enter states of empathy, as we heard earlier. So, careful prior planning of how we spend our time and what we eat will give us a head start in creating the right conditions for psychological flourishing.

I think of this as analogous to gardening. We are so used in the modern world to rushing around trying to achieve things by sheer strength of will, yet gardening is about creating the right conditions and then being patient and letting nature take the lead. So too, by making time for meditation and contemplation, by balancing our rest, work and exercise, by eating well, we best prepare the soil for our own mental transformation. We cannot simply command it on the spot, like we might order a parcel on Amazon at the click of a button. Happiness is something that slips in on us unnoticed when the conditions are right. In the modern world, there are so many negative influences, like pernicious garden weeds, threatening to overrun our minds and bodies and prevent positive states of mind. Junk food advertisements influence our diet, affecting the beneficial bacteria in our guts, and junk media interferes with the reflective thoughts in our minds that are conducive to self-transformation. So we must ensure that we feed our inner ecosystem the right balance of nutrients, while feeding our brains selectively too. As meditation neuroscientists Daniel Goleman and Richard Davidson claim: 'a haphazard mental diet most likely leads to equally haphazard changes in the muscle of the mind'. It pays to think about what you are feeding your brain as much as your stomach.

We all know habits are hard to change, but if we can find our own best way to build good habits, we can take positive action almost mechanically without the need for conscious effort. The good thing about this approach is it allows strategic planning. With foresight, we can design a garden where the flowers of enlightened self-consciousness are more likely to bloom.

These are just a few but there are many tools we can employ to help us better embrace other people and the natural world as part of our self-identity. We will certainly need them, because it can be very difficult, especially in our busy, media-soaked world, to avoid getting drawn back into an individualistic mindset. Many aspects of the broader environment influencing our body and minds are out of our control, such as the design of the towns we live in, the quality of the air and the availability of nutritious food. Such food may be unaffordable, or we may live or work in a 'food desert' where healthy options are less accessible. We also have limits on how we spend our time – we may have a very time-intensive job or family and caring duties. Furthermore, some of the influences of our environment affecting our self-identity are very subtle, such as the way people talk and the words they use. Our brains are products of our culture and like sponges they absorb the language and discourses they are immersed in. The writer David Foster

Wallace once told a joke about two young fish. They happen to meet an older fish swimming the other way, who nods at them and says, 'Morning, boys, how's the water?' The two young fish swim on for a bit, and then eventually one of them looks over at the other and says, 'What the hell is water?'

For humans, language is our water and as the writer James Carroll said: 'we swim in language, think in language, we live in language.' And so what happens when that language changes imperceptibly, like the increasing use of individualist words in literature, news and pop songs occurring over time?[10] The result is that the way our brains work also changes imperceptibly, and such changes are not always in a random or benign direction. Our culture and social context can be steered by those with vested interests. Companies and governments often try to do just this. As the economist Tim Jackson writes:

> Government policies send important signals to consumers about institutional goals and national priorities. They indicate in sometimes subtle but very powerful ways the kinds of behaviours that are rewarded in society, the kinds of attitudes that are valued, the goals and aspirations that are regarded as appropriate, what success means and the worldview under which consumers are expected to act. Policy signals have a major influence on social norms, ethical codes and cultural expectations.[11]

Sometimes government influences are very obvious, such as when British Prime Minister Margaret Thatcher boldly proclaimed 'there's no such thing as society. There are individual men and women and there are families.' This is a less than subtle nudge towards an individualistic outlook. Combine this 'social modelling' with the narcissism-inducing advertising we are daily bombarded with by corporations – 'Because you're worth it' (L'Oreal); 'You deserve a break today' (Mc-Donald's) – and is it any wonder that our culture, and the minds it moulds, have tended towards greater individualism? Research has shown that exposure to materialistic messages causes people to adopt materialistic and self-interested values themselves.[12] Going back to our imaginary graph with two axes related to psychological and material connectivity, it is perhaps not surprising that in most countries we have seen modern cultures evolve into the quadrant characterised by psychological isolation while material connectivity has increased.[13]

However, despite the prevailing current of greater individualism across the world, things are often more complex than they seem on the surface. If one narrative becomes dominant in a culture at a given point in history, there are still many other competing narratives under the surface. When cultures clash, our minds are the battlegrounds those wars play out in. And there is certainly a battle going on in the twenty-first century. On the one hand, there is a growing awareness of the illusion of our self-identity – ancient insights based on introspection and intuitive reasoning are increasingly supported by scientific evidence from microbiology to neuroscience, from ecology to psychology. These scientific strands are being drawn together to outline a more comprehensive scientific understanding of our intimate biophysical and psychological interconnectedness to the world around us. The consequences of remaining stuck in a state of isolated self delusion are also becoming painfully clear – from personal loneliness, depression and other mental health problems to broader social and ecological impacts in the wider world resulting from our selfish tendencies. In his book *Homo Deus*, Yuval Noah Harari writes that we have known for thousands of years the self is an illusion but that such doubts 'don't really change history much unless they have practical impacts on economics, politics and everyday life'.[14] Yet, now we know there are indeed clear practical impacts of our limited understanding of the self: if we cannot shed the self delusion and act with appropriate concern for others and the world around us, then we will face ecological and climate catastrophe, which has far-reaching implications for economics, politics and our everyday lives.

At the same time as this rapidly emerging awareness of the personal and planetary implications of self-identity, we also face a powerful opposing culture of individualism. The overall trend in many countries is still towards *greater* individualism, if not downright narcissism, and this seems set to continue for the near future. There are powerful vested interests influencing our mainstream media, pushing a right-wing agenda that has advanced nationalism and protectionism as an acceptable viewpoint. And there seems to be no let-up in product advertising that promotes individualism (it's no wonder really, because it pays off for advertisers – we are simply more likely to buy things for ourselves when we are made to feel selfish).

There is certainly a cultural battle to be faced in trying to transform minds to a broader ecological and social self-awareness. Unfortunately, there is also some urgency. The clock is ticking because we are likely approaching an earth-system tipping point where we will soon become unable to reverse the impacts of climate change.[15] Similarly, our chemical pollution of the biosphere and the rapid destruction of the world's biodiversity may be similarly difficult, if not impossible, to reverse.[16] What's more, other environmental shocks are set to become worse in the future, such as sea-level rise, antibiotic resistance, extensive wildfires, crop failures, droughts and powerful storms. The mass human migration these will cause will likely lead to a knee-jerk defensive reaction in society. Under stress, the natural response of human societies is to focus on protecting a smaller in-group. From historical studies, cultures have been shown to become 'tighter' in the face of environmental shocks – they become more collective and cooperative internally but develop greater hostility to out-groups. A global analysis in 2011 found that countries that have experienced more ecological and social shocks over their history (e.g. resource scarcity, territorial conflict, and disease and environmental threats) had 'tighter' cultures, meaning they have strong social norms and rules, allowing close internal cooperation, but these cultures also developed greater hostility to out-groups, towards whom they acted less favourably or even antagonistically.[17] We perhaps see the start of a new trend in rising nationalist sentiment across many Western democracies. There seems to be increased hostility towards people migrating into countries, and towards diaspora who have been settled for decades. We may think the number of people on the move across the globe right now is substantial (258 million people on the move in 2017 for example), but this is just a trickle compared with how it will be when climate change gathers pace.[18] How will levels of xenophobia and nationalism evolve in response to this?

At the start of the new millennium, the world conducted its first global 'ecosystem assessment', part of which involved developing scenarios for how our society and environment might evolve. One of those scenarios was 'fortress world' where rich countries build barriers to keep migrants and refugees from poorer countries out. The barriers may be physical (think thousands of people chanting 'Build a Wall' at Trump rallies to keep Mexicans out of America), and they may also be psychological – the devaluing of people from other cultures by treating them as out-groups. This dark road outlined by the Millennium Ecosystem Assessment unfortunately seems to be one we are headed down, and it is an ethical time bomb. Rich countries are causing global climate and ecological catastrophe and then building walls to prevent millions of people moving out of inhospitable conditions. In the face of malnutrition and deaths that are ultimately caused by the actions of richer countries, the only way to make such news palatable will be to devalue and dehumanise the people outside the fortress walls – for a start, referring to them as 'migrants' or 'criminals' rather than 'refugees' or simply 'people' (as we have learned, it is easier to blame when we ignore people's context and allocate them broad characteristics). The news media will explain that it is the failure of their governments causing these problems, even though governing under such inhospitable environmental conditions will be a near impossible task. At some deep level, however, we will still know it is us in richer countries who are responsible, and a feeling of sickly guilt will permeate our culture. Will that be enough to prompt us to change our ways? Perhaps not – it has been shown that under conditions of psychological insecurity there is actually an orientation to *more* materialistic values.[19] As people begin to worry such things might happen to them, and as they *do* start to happen to them – because even rich countries will not be truly immune to environmental shocks, such as extreme weather events, but only partly buffered by their wealth – then we may become even more selfish. While Rome burns, we will go on a shopping spree.

There is a window of opportunity to change our mindsets now. The window is still open, but unless we take the chance to reform our self-identity and change the way we interact with each other and the natural world, we are set to continue a global 'race to the bottom'. The challenge is stark: we need to fix our national and global institutions to better manage the climate, pollution, habitat loss, animal welfare and social degradation, and the only way to effectively do this is to tackle the root cause – to change our personal mindsets. Although we may be strongly motivated, however, we must remain ever cautious because the social context we are in also heavily influences us. We must somehow devise ways to inoculate society against the knee-jerk reactions that will come as environmental shocks gets worse. Closing ranks when adversity strikes is a natural and adaptive group response that has allowed humankind to overcome hard times during its evolutionary past. But as the global population increases further, taking an inward-looking perspective to protect the in-group will only lead to *greater* instability, because we ignore the costs that accumulate on other out-groups and the environment, and on a finite planet these will simply come back and bite us. We need to overcome an instinctive reaction that was once adaptive and has now become maladaptive in the modern world. Just as we have learnt in earlier chapters how a sense of discrete, autonomous self-identity evolved to benefit survival but has now become maladaptive, so too the natural social reaction to protect the in-group in times of adversity is a cultural adaption that has become maladaptive in the modern world. If we want to thrive and pass on a habitable world to future generations, we must overcome both our biologically determined egoism and our culturally determined tribalism. We must overcome both the 'selfish gene' and also the 'selfish meme'. If we can, we just might be able to steer cultural evolution into the quadrant where we achieve both psychological and material connectivity, and we just might be able to secure a stable and equitable environment for humanity.

19

Joining the dots

Look outside the window at a garden or a patch of green space, see the grasses and shrubs and trees. Where are the boundaries between these things and how real are they? Take what looks like a single plant. Inside its leaves, chemical reactions are buzzing away converting sunlight into sugars. As carbon dioxide enters through myriad pores, oxygen molecules stream outwards. Below ground, a web of root tendrils are working cooperatively with the filaments of mycorrhizal fungi to draw chemical nutrients from the soil. A host of organisms, such as small sap-sucking insects, leaf miners, bacteria, snails, squirrels, deer, will connect with this plant to extract the energy it derives from the sun and soil. It is also in communication with its neighbours – a damaged plant releases volatile chemicals that other plants detect and use to upregulate their own defences in near real-time social alerts. What appears to be a distinct entity is actually a dense confluence of connections in an ecological network: an eco-system.

We humans too have ecosystems inside of us and, like Russian dolls, we are integral parts of broader ecosystems, connected to the flora and fauna around us. Like the plants and their roots, our bodies are open systems – energy flows through our cells, which are comprised of matter once part of other animals and plants. The pulse of cells in our bodies is part of a continuum of hundreds of trillions of cells across the world, all sharing minor variations in the same DNA code. Non-human organisms comprise a substantial part of our bodies, colonising nearly every surface of our guts, skin and brain, and constantly sharing material and energy with us.

The current bundle of cells we call our body also forms a mind that is closely linked with other minds. We transfer electrical messages fluidly between neural networks every time we communicate. The airwaves around us buzz with conversations encoded in both sound and electromagnetic waves of energy. At the larger scale, air routes, shipping lanes, roads and electricity grids connect us across the globe, like the nervous system and arteries of some great superorganism, allowing a continual flow of information and material between us.

Through impressive scientific advances we have learnt in a reductionist way about the different parts of our world and ourselves. Now, increasingly, the focus is on the broader systems in which these parts fit together. It is an exciting time for modern science. Social scientists Nicholas Cristakis and James Fowler describe the advent of this broader assembly project. They highlight how over recent centuries scientists were swept up in a 'reductionist fervour', examining ever-smaller bits of nature: biologists studying human organs, then cells, then molecules and genes; physicists breaking down matter into atoms, then nuclei, then subatomic particles. Now, though, across many disciplines, scientists are trying to put the parts back together:

> whether macromolecules into cells, neurons into brains, species into ecosystems, nutrients into foods, or people into networks. Scientists are also increasingly seeing events like earthquakes, forest fires, species extinctions, climate change, heartbeats, revolutions, and market crashes as bursts of activity in a larger system, intelligible only when studied in the context of many examples of the same phenomenon. They are turning their attention to how and why the parts fit together and to the rules that govern interconnection and coherence.[1]

Such 'systems-science' is highly relevant in our endeavour to understand ourselves. Our human condition is a systemic one – we comprise many systems linked across different dimensions: physical, mental and social. Our body is an emergent property of trillions of cells functioning together and interwoven with the world around us. Although each individual cell lives but a short while, the design that governs our body structure is encoded in long-lived DNA instructions borrowed from our ancestors and which we will pass on to generations to come. Our mind too is an emergent property arising from continual loops of electrical stimulation in our brain's neuronal circuits.

Your mind is connected intimately to others, often sharing information between connectomes. The sum of all the minds and bodies currently alive make up our social system. In this emergent social structure, roles are specialised just like cells in the human body. Individual humans can come and go but the bakers, builders, clerks, waste technicians, firefighters and police that society needs to function all form part of a structure which remains intact and outlives the temporary parts. The actions of societies lead to the emergent property of the global economy, which in turn impacts the ecology of the living world around us. Our world is one of deeply connected interacting systems.

To understand our human identity, we should not therefore attempt to draw a boundary too small – let us not believe that our selves simply stop at the boundaries of our skin. The most defining feature of humanity is the great level of connectivity and cooperation between us, both over global extents and long time spans, which is facilitated by our culture, yet we are so often ignorant of these processes, because they occur at scales we cannot directly perceive. We are used to viewing the world from behind a pair of eyes that see to only a limited spatial extent, we interpret events in terms of their relationship to an illusory self that lasts just the lifetime of a single body. However, there is no fundamental reason that the spatial and temporal scale at which we commonly perceive the world is any more real than other scales. As evolutionary biologist Scott Sampson writes: 'One of the greatest obstacles confronting science education is the fact that the bulk of the universe exists at either extremely large scales (e.g. planets, stars, galaxies) or extremely small scales (atoms, cells, genes) well beyond the comprehension of our (unaided) senses.'[2] We are blind to the very slow or the very fast processes that connect us, blind to the processes that occur at microscales or planetary levels and which bind us together. Our limited perspective means we are often trapped by seeing ourselves falsely at the centre of a small world. We suffer from a problem similar to the heliocentrism of earlier generations (the belief that the earth was at the centre of the universe); we suffer from egocentrism – the illusory belief that we are a coherent self at the centre of a universe. Yet, just as we have learned to accept the non-intuitive fact that the Earth is not at the centre of the solar system, so too must we now accept that we ourselves are not the centre of the universe, but that we are fully enmeshed within it.

Of course, altering our perception is likely to be difficult, because our individual point of view is programmed into us, both by biology and by our culture – by our mainstream education and the broader media within the increasingly dominant Westernised culture. Yet, this highly atomised, individualistic perspective is becoming maladaptive, leaving many people feeling increasingly isolated and suffering from mental health problems. It underlies an *individuation pathology*, in which an ill-informed and limited perspective on the world negatively impacts other humans and non-humans around us. Just like how a butterfly flapping its wings could affect the weather on the other side of the world, so too do the consequences of our actions cascade across the surface of the globe and, over time, across multiple generations. Our mindsets of illusory independence determine the structure and functioning of our institutions, and at the moment these are acting like great bulldozers that destroy the natural world with little moral oversight or restraint. The consequences of all our actions accumulate to become vast tidal waves that destroy the natural and social basis that underpins our prosperity. If we want to solve these problems, and we urgently must, we need to find a way to break free of the subjective delusion of self-isolation and embrace a greater responsibility for our impacts on the world. We need a more encompassing sense of self-identity that changes the way we relate to the world.

There are many ways to transform our relationship to the world around us. Numerous philosophers, writers and scientists have suggested methods such as changing our language, practising meditation, engaging in outdoor education, or taking psychoactive drugs. We have touched on some of these, and I do not advocate which would work best (it probably depends very much on your context). However, some caution is needed, because what is required is a subtle shift in emphasis, not a wholesale abandonment of our individual minds, which would prevent us functioning successfully in the world. If reading to seek deeper knowledge can help steer our worldview towards a balanced self-identity, then this book might hopefully have some small role in that transformation. It is likely, however, that more practical approaches will also be needed to counter the strong sway of both our biological tendencies and prevailing individualistic cultures, in order to help our connected sense of self become more dominant.

We will need to remain on guard because our minds have a tendency to revert, like the Necker cube illusion, back to a viewpoint where the world revolves around a limited and isolated sense of self. It can be difficult to escape the mental cockpit of a fully autonomous 'little me' steering our bodies through the world, yet the growing body of science across different domains stacks up against such an interpretation.

Although we sometimes have brief and beautiful glimpses beneath the veil of delusion – we momentarily feel we are in a world where we are all deeply equal and interconnected with each other – these meaningful insights are rarely permanent. So how can we use our objective understanding to repair our innermost belief system and ensure it is built on a firm and true footing?

In systems science, there have been advances in understanding the way in which social, economic or ecological systems can get 'locked' in certain states which are undesirable for society, and this research may have relevance for understanding the persistence of brain functional states too. Consider an ecosystem for a moment – a beautiful reef in the Pacific Ocean full of colourful fish and coral. In the face of pollution and overfishing, reefs can rapidly switch from being thriving species-rich habitat to being dominated by just a very few creatures, such as spiny sea urchins or starfish. In the face of pressures on the ecosystem, biological communities rapidly pass a tipping point causing them to 'flip' into an alternative degraded state. Returning them to their previous condition can be very difficult, even when environmental pressures are reduced.[3]

My own research has explored the extension of these systems concepts to broader social and economic systems, such as our global food system. The way we have developed the production, manufacture, distribution and retail of food is quite efficient in a narrow sense, in that it allows many of us to reliably purchase diverse types of food grown from across the globe (although many people still lose out, with millions of people still suffering from malnutrition). However, the global food system also has extensive negative impacts on climate, biodiversity, water quality and human health. With a broader team of interdisciplinary scientists, I have tried to understand which factors constrain the transformation of the food system away from a state with negative impacts. Some of the mechanisms that 'lock in' the system are economic (for example, agricultural subsidies), some are biophysical (the destruction of species that control pests, meaning we are stuck to using chemical pesticides), while others reside within our minds (a tendency to believe in 'technofixes' – the belief that we can solve environmental crises simply through future technological solutions).[4] Other social and environmental 'lock-ins' such as poverty traps, invasive species, diseases and runaway greenhouse gas emissions can be explored in a similar way – by unpicking the mechanisms that prevent us from steering social, economic and environmental systems towards a more sustainable course.

What relevance does this have to our discussion of self-identity? Well, perhaps in a similar way that these broader social and environmental systems are 'locked in' to negative states, so too are our minds 'locked in' to an individualistic state. Could we explore the different psychological, social and cultural mechanisms which trap us in these mindsets, thereby allowing us to free ourselves from self delusion?

Psychologists and medical scientists are already starting to take note of the idea of neural systems in our brains flipping between alternative states. It has important relevance to conditions such as epilepsy and depression. Researchers have explored whether there might be 'early warnings signals' before people transition between neurological states; for example, the brain waves of epileptic patients show measurable changes in electrical patterns before an oncoming seizure. These early warning signs can be detected with devices such as portable EEG headsets, mattress sensors or watches detecting subtle body movement.[5] Similarly, psychologists have found that patients susceptible to becoming trapped in long-term depression first begin to 'flicker' between depressed and non-depressed states, allowing preventative action to be taken early on.

Perhaps the transition from an isolated sense of self to a more expansive, and accurate, sense of selfhood will show the same kind of nonlinear 'tipping point' dynamics as these other social and environmental systems described above. It is just speculation at this stage, but there is emerging evidence that experienced meditators have long-term persistent differences in their brain functioning compared to non-meditators. Meditators experience more compassion, as well as seeing themselves as being more integrated with the rest of the world, and they seem to have settled in this brain state more permanently. People who meditate, even for twenty minutes per day, begin to show changes in brain functioning, although the new state is fragile to perturbation, such as stressful events.[6] This flickering between states, between 'little me' and 'big me' – from an isolated to a more expansive and networked sense of self-identity – is encouraging. It suggests that these minds may be on the cusp of a more permanent state shift. A tipping point in self-identity may be approaching.

In an almost fractal-like way, these changes in individual

minds then influence a tipping point across multiple minds in our connected society. The writer Malcolm Gladwell outlines how trends and behaviours, such as high crime rates, spread unexpectedly rapidly through human populations.[7] These phenomena reach a critical mass in a small subset of the population before a tipping point is reached and they spread like wildfire across the population. Might the same one day occur with regards to our sense of self-identity? As the mindset of each person changes, it makes influencing the transition in others more likely. Could we use such social contagion to enable the rapid evolution of a networked sense of self-identity among the global human population?

It is a challenging, and daunting, task perhaps. It can be unnerving to think about the root cause of social and environmental problems as lying within our individual minds. We like to think about proximate solutions – solve climate change by taxing carbon, solve food shortages by the genetic modification of plants and animals, solve mental health problems by building green infrastructure in our cities. These technical fixes have their place, but we cannot ignore the part played by our individual belief systems that, cumulatively across populations, underpin our global institutions.

Is changing the mindset of nearly 8 billion humans on the planet in time to solve pressing global sustainability problems a feasible task? When we look at how quickly behaviours and attitudes spread once a tipping point is reached, I feel hopeful that such a transition could occur, and with globalised digital communication networks making us more connected than ever before, social transformations can occur faster than at any time in history. Change will be hard for some people, because their neural networks have become more fixed, but young people and future generations to come will no doubt think differently to us. Growing up with a greater awareness of problems like climate change and biodiversity loss, people will be much more motivated to solve these problems, and they may be more savvy as to how their minds can be manipulated – how branding by companies and fake news influences their self-identity. Becoming immune to these manipulations and helping promote a sustainable society by becoming less selfish might eventually become the obvious choice. For the rest of us, whose minds are set in their ways, we can take solace in the knowledge that change is possible – human brains are highly plastic and can be modified with training. Every moment we operate in the expanded mode of self-identity, it makes catalysing positive change in ourselves and others more likely.

As any system moves towards a tipping point, the conversion of every single node in the network becomes critical. As your sense of self-identity widens and your attention lifts, from perceiving yourself as a single thread to seeing the majesty of the whole tapestry, you become a critical part of the global effort towards securing a safer and happier future for all humanity.

Acknowledgements

In a famous essay, the economist Leonard Read describes the thousands of people that contribute to the design and manufacture of a simple object such as a pencil ('my antecedents are legion'). So it is with this book: hundreds of thousands of people contributed the knowledge synthesised here, and if it wasn't me writing it, it would surely have been someone else. Inventions are usually an inevitable next step in a long chain of interdependent innovations, and so we should give proper due to our antecedents for the inexorable roll of progress. I therefore humbly acknowledge the legions that helped to create this book. That said, there are a few more proximate catalysts who should warrant specific acknowledgement. I thank my fantastic and dynamic agent Jen Christie for believing in this project. She showed new ways to present ideas and greatly improved my writing, at times helpfully dragging me away from the tight academic prose I was used to. I also thank my editors Jenny Lord, Paul Murphy, and other colleagues at W&N for their professional guidance in honing the content, and colleagues at the University of Reading, especially Pete Castle, for their professional support. My family and I shared useful discussions around ideas in the book, along with Eric Allan, Matt Heard, Daniel Kelly, Mark Pagel, Amy Proal and Miles Richardson.

I thank Lancaster library for allowing me as a young child to explore their books and foster an interest in ancient eastern religion. The closure of many libraries in the UK really is a false economy with regards to social progress. Finally, my thanks to colleagues at the European Environment Agency and Defra for immersing me in ideas about interconnected global social and environmental issues.

Further Reading

I have drawn upon a wide range of materials and shall not attempt to provide a full bibliography of primary research papers here – although some of these can be found in the *Notes* section. I detail sources which have been influential to me and provide the reader with context to the ideas in this book.

For general texts on the 'systems perspective' I recommend the excellent work of Fritjof Capra (especially *The Turning Point*, 1983, Flamingo, London, and *The Hidden Connections*, 2003, Flamingo, London) and Frederic Vester's *The Art of Interconnected Thinking* (Mcb Verlag, Munich, 2012). For the theme of *Our Interconnected Bodies*, Sebastian Seung's *Connectome* (Penguin Books, London, 2012) is a good introduction to the links between the billions of nerve cells in our bodies, while a book called *I, Superorganism* by Jon Turney (Icon Books, London, 2015) describes the connections between human cells and our non-human microbiome. *The Ancestor's Tale: A Pilgrimage to the Dawn of Life* by Richard Dawkins (Weidenfeld & Nicolson, London, 2005) beautifully describes how DNA flows through different plant and animal forms to weave the web of life. For the theme of *Our Interconnected Minds*, I recommend texts such as *Connected: The Surprising Power of Our Social Networks and How They Shape Our Lives*

(Little, Brown, New York, 2009) by Nicholas Christakis and James Fowler. For better understanding *Our Self Delusion* a good start would be Buddhist texts, for example, *The Art of Living*, by Thich Nhat Hanh (Rider, London, 2017), popular neuroscience books like *Social: Why Our Brains Are Wired to Connect* by Matthew Lieberman (Oxford University Press, Oxford, 2013), and psychology texts such as Richard Nisbett's *The Geography of Thought* (Nicholas Brealey, London, 2003). Building from these to think about a more accurate systems perspective on the human condition (*Our Network Identity*) and its attendant benefits for mental health, one might start with *The Science of Meditation* by Daniel Goleman and Richard Davidson (Penguin Random House, London, 2017) as well as initiatives such as the Center for Healthy Minds (https://centerhealthyminds.org/). In terms of the wider social and environmental benefits that emerge from revising our self identity, I recommend *The Ecology of Wisdom* by Arne Naess (Penguin Random House, London, 2008), *Ecopsychology – Restoring the Earth, Healing the Mind* (Counterpoint, Berkeley, 1995) and a World Wildlife Fund report by Tom Crompton and Tim Kasser, *Meeting Environmental Challenges: The Role of Human Identity* (WWF-UK, Surrey, ISBN: 9781900322645, 2009). In terms of source material for my research, I used several magazines that acted as landing points for a deeper dive into the primary literature. To this end, *New Scientist* and *Resurgence* magazine have been helpful to me, as also have the webpages of Wikipedia (https://www.wikipedia.org). As an academic professor, I would probably not suggest my students use it is a primary source for citation, but as a starting point to explore new subjects, following up links within the pages, it is exceptionally helpful (and surprisingly accurate through mutual checking from more than 2.5 million editors).[1] It is a wonderful testament to how millions of human minds from around the globe are working in concert to create a dynamic body of knowledge that no individual mind could encompass alone. Science is about providing the evidence to see the world in a more accurate way, but how we act on that evidence to change ourselves depends on values and choices. Fiction helps us to understand these, and to this end I recommend two books which frame the potential outcomes of a world where individuality is taken to the extreme (*The Road* by Cormac McCarthy, Picador, London, 2006) or abolished almost absolutely (*We* by Yevgeny Zamyatin, Penguin, London, 1993).

Notes

Introduction

1   This quote is disputed, and most commonly attributed to
    economists John Maynard Keynes (1883–1946) and Paul
    Samuelson (1915–2009).

Chapter 1

1   R. M. Forbes, A. R. Cooper and H. H. Mitchell, *J. Biol. Chem*
    (1953). Admittedly, I may have cooked up the storm for dramatic
    effect and left a few of the more technical sentences out.
2   This is the global average in 2005. Developed regions have far
    higher averages (e.g. Europe 70.8 kg, N. America 80.7 kg). See
    S. C. Walpole et al., *BMC Public Health* (2012) 12, 439.
3   Calculated using the radius of the earth as 6,371km from the
    NASA earth fact sheet (http://nssdc.gsfc.nasa.gov/planetary/
    factsheet/earthfact.html) then using the equation for the volume
    of a sphere $4/3.\pi.r^3$. This process is then repeated with a larger
    sphere of 6,471 km radius (including the 100 km extra radius up
    to the Karman line), followed by subtracting one volume from
    the other. Note, this ignores mountains and valleys below sea
    level, so it's a rough guess!
4   Fortunately many people have thought long and hard about
    how densely spheres can be packed into spaces (see https://
    en.wikipedia.org/wiki/Close-packing_of_equal_spheres). In
    this case we know the number of molecules and the volume, so

we can try to find the largest sphere of space possible that would surround each molecule if they were spread evenly across the volume.

5 Supernovae occur during the last life stages of a massive star. The star's destruction is marked by one final huge explosion, which causes the star to become hugely brighter when observed from afar and slowly fading over time. Anglés-Alcázar et al. (2017) *Monthly Notices of the Royal Astronomical Society* 470, 4698–4719

6 F. Capra, *Hidden Connections* (Flamingo, London, 2003), p. 59.

Chapter 2

1 Los Angeles Times archive April 7, 1994 https://www.latimes.com/archives/la-xpm-1994-04-07-mn-43323-story.html.

2 The new management team was led by Steve Bannon, the former executive chairman of Breitbart News and White House Chief Strategist for U.S. President Donald Trump.

3 Donella Meadows archives http://www.donellameadows.org/archives/biosphere-2-teaches-us-another-lesson.

4 The animals in the first Biosphere 2 mission included four African pygmy goats, thirty-five chickens, three dwarf pigs as well as tilapia fish in the pond system. J. Allen and A. G. E. Blake, *Biosphere 2: The Human Experiment* (Viking, New York, 1991).

5 S. E. Silverstone and M. Nelson, *Advances in Space Research* (1996) 18, 49–61.

6 D. V. Spracklen et al., *Nature* (2012) 489, 282–285.

7 Based on daily consumption of around 1.2 litres of water, which is the amount needed to stay hydrated according to the UK National Health Service (although obviously this varies between individuals and is greater for those living in hotter places), and 1.35 kg of food, with a lifespan of 71 years (the global average in 2010–15 according to the United Nations *World Population Prospects* – 2017 Revision).

8   Assuming a tidal volume (average breath) of 500ml and a resting
    breathing rate of around 16 breaths per minute (about average)
    with a lifespan of 71 years. Add more if you exercise regularly!

9   S. Törnroth-Horsefield and R. Neutze, *PNAS* (2008) 105,
    19565–19566.

10 https://www.timeshighereducation.com/news/
    life-span-of-human-cells-defined-most-cells-are-younger-than-
    the-individual/198208.article.

Chapter 3

1   There are, as always, exceptions to the rule: the skin condition
    Rosacea is often associated with numbers of *Demodex* mites
    around 15–18 times higher than in healthy subjects, and they may
    well be a causal factor in the condition. S. Jarmuda et al. *Med
    Microbiol* (2012) 61, 1504–1510.

2   R. Sender et al., *PLoS Biology* (2016) 14, e1002533.

3   S. J. Song et al., *Elife* (2013) 2, e00458.

4   The bacterium *Staphylococcus aureus* aggravates symptoms of
    atopic dermatitis, but its ability to colonise our skin is limited
    by a diversity of other 'friendly' skin bacteria. T. Nakatsuji et al.,
    Science Translational Medicine (2017), 9, eaah4680

5   There is some debate as to whether the reduction in the
    microbiome is a cause or consequence of these diseases. A. Mosca
    et al., *Front. Microbiol* (2016) 7, 455 cites evidence suggesting that
    it may be the former in many cases because a loss of microbial
    diversity in the gut often precedes these diseases.

6   This loss of a genes' function can occur by random genetic 'drift'
    whereby a mutation occurs in the gene. Because the symbiont
    is producing a similar product, the gene's function is no longer
    essential and the host organism's reproductive success is
    unaffected. In this case, the non-functional gene is passed on to
    future generations. In other cases, the gene may be lost completely
    because keeping it is costly (e.g. producing a redundant product)

and therefore natural selection favours individual organisms in which the gene is down-regulated or lost. J. J. Morris et al., *mBio* (2012) 3.

7   Red blood cells expel their organelles, including mitcochondria, during cell maturation. For their role in ferrying oxygen around the body, they need very little energy, and this leaves more space for haemoglobin to transport the oxygen.

8   M. W. Gray et al., *Genome Biology* (2001) 2, PMC138944.

9   H. K. Kim et al., *Cell Metabolism* (2018) 28, 516–524.

Chapter 4

1   The gene is called the *egt* gene and disrupts a moulting hormone that normally causes the caterpillar to climb downwards. K. Hoover et al., *Science* (2011) 333, 1401.

2   The beaver dams are an example of extended phenotype given by Richard Dawkins, although I prefer the caddis fly example for which learning behaviour can be completely discounted. In organisms with more developed nervous systems and greater social interactions, behaviours to manipulate the environment could be genetically encoded but they may equally be learned through culture. This is the case with human buildings of course – there are no genes for skyscraper building, but rather a great mass of engineering knowledge that we pass across generations by word of mouth and the written word.

3   Usually after around 2–10 days following the onset of symptoms, with survival rates of less than 10 per cent even with treatment.

4   S. Gluska et al., *PLoS Pathogens* (2014), 10, e1004348.

5   In a particularly poetic analogy (perhaps not something you would expect from a scientist studying viruses) Professor Warner Greene from the Gladstone Institutes in San Francisco describes this process whereby immune cells are recruited but then killed: '*The cavalry come riding in and fall victim to this same form of fiery cell death, turning their rifles on themselves.*' This is from an

interview with *The Scientist* magazine in which Professor Greene describes two seminal papers advancing the understanding of the interaction between HIV virus and the immune system through the body's inflammatory response. It can be found in full here: http://mobile.the-scientist.com/article/38739/how-hiv-destroys-immune-cells.

6   Infection rates vary by country, with studies in some countries showing up to 90 per cent infection rate while others are as low as 10 per cent. Infection rates tend to be higher in more humid countries and in older people. J. Flegr et al., *PLoS ONE* (2014) 9, e90203.

7   J. P. Webster, *Schizophrenia Bulletin* (2007) 33, 752–756.

8   Although in a very speculative mood one might think about how on the Great Plains, where humans evolved, big cats may have been a significant cause of mortality. So could there have been benefits in avoiding signs of cats, which *Toxoplasma* would benefit from overriding?

9   There are a number of other *Toxoplasma* effects in humans that show these sex-specific differences, although the reasons are not yet fully understood. See J. Flegr et al., *PLoS Neglected Tropical Diseases* (2011) 5, e1389.

10 J. Flegr, *Journal of Experimental Biology* (2013) 216, 127–133.

11 You can read more about this experiment exploring the remarkable fluid dynamics of snot during a sneezing event at: http://www.nature.com/news/the-snot-spattered-experiments-that-show-how-far-sneezes-really-spread-1.19996.

12 Sun et al., *PLoS Biology* (2019) 17, e3000044.

13 From M. Horie et al., *Nature* (2010) 463, 84–87. Some of this DNA codes for latent retroviruses with the potential still to function, but much of it is fragments of old viruses forming part of the junk DNA that biologist Richard Dawkins likens to obsolete files cluttering up our DNA 'hard drives'.

14 D. Graur, *Genome Biology and Evolution* (2017) 9, 1880–1885.

15 J. F. Brookfield, *Nature Reviews Genetics* (2005) 6, 128–136.

Chapter 5

1   A DNA nucleotide consists of a phosphate group, a five-carbon sugar (deoxyribose) and one of four nitrogenous bases (thymine, adenine, guanine or cytosine), which are organic molecules with a nitrogen atom and the chemical properties of a base.

2   It is fascinating to consider the many media across which this code traverses to reach our senses. The binary code of the Beethoven symphony is first encoded digitally in a computer chip, then sent across airwaves as electromagnetic patterns in a Wi-Fi signal, then it is transmitted back into electric charges in the receiving computer, and then through a speaker system which causes physical compression of air reflecting the code, which is received as sound by our ears. The journey doesn't end there. The nerves in the ear transmit the code into electric signals in nerve cells which are then interpreted by various parts of the brain- deriving emotion and meaning from the signals. These complex responses are only possible from the elegant simplicity of the code allowing it to be so easily and faithfully transmitted.

3   Each cell differs in its DNA by around 100 base pairs, estimated by applying the mutation rate of one in every 30 million base pairs copied to the roughly 3 billion base pairs of the human genome which is copied each time a cell divides.

4   J. Turney, *I, Superorganism* (Icon Books, London, 2015).

5   R. Dawkins, *The Ancestor's Tale: A Pilgrimage to the Dawn of Life* (Weidenfeld and Nicolson, London, 2005).

6   Although admittedly it does affect the *context* of the existing information units, which can affect the way they are expressed in the phenotype of the individual. So, for example, a recessive gene is expressed differently in combination with another recessive gene versus with a dominant gene.

7   N. A. Moran and T. Jarvik, *Science* (2010) 328, 624–627.

8   J. L. Blanchard and M. Lynch, *Trends in Genetics* (2000) 16, 315–320.

9   O. G. Berg and C. G. Kurland, *Molecular Biology and Evolution* (2000) 17, 951–961.

10 For the original discovery of these virus '*syncytin*' genes, which in human placenta development help to form cells joining the placenta to the uterus and drawing nutrients from mother to baby, see S. Mi et al. *Nature* (2000) 403, 785. Since then, the gene has been found in a range of other mammals from other primates to cats, dogs and mice, suggesting transfer from the virus to an early shared common ancestor of all these animals and then conservation through time. More recently, another gene carrying out a similar function but from a *different* virus has been found in rabbits, showing just how common these horizontal gene transfer events can be, especially when a virus produces something that a host organism can benefit from. O. Heidmann et al., *Retrovirology* (2009) 6, 107.

11 A. Crisp et al., *Genome Biology* (2015) 16, 1–13.

12 C. R. Woese, *Microbiology and Molecular Biology Reviews* (2004) 68, 173–186. The discovery of horizontal gene transfer led *New Scientist* magazine, on the 150th anniversary of Darwin's *Origin of the Species*, to declare on their front cover, somewhat sensationally, 'Darwin was Wrong'. To be fair, he got an awful lot right, yet it would need another century of research with new molecular techniques to see how genes are transferred not only down, but also (albeit much less frequently in higher organisms) across, generations.

13 Endosymbiotic theory suggests that organelles, such as mitochondria, in eukaryotic cells originate from larger host cells enveloping and integrating previously free-living prokaryotic cells.

14 L. Margulis and D. Sagan, *What is Life?* (University of California Press, Berkeley, 1995).

Chapter 6

1   That is around 15 million new connections made every minute. Of course, many connections are also lost, and it is this plasticity of the brain which allows learning. B. Hood, *The Self Illusion* (Constable, London, 2011).

2   S. Seung, *Connectome* (Penguin Books, London, 2012).

3   Donald Hebbs' theory (now referred to as Hebbian theory) is as follows: 'Let us assume that the persistence or repetition of a reverberatory activity (or "trace") tends to induce lasting cellular changes that add to its stability. . . . When an axon of cell A is near enough to excite a cell B and repeatedly or persistently takes part in firing it, some growth process or metabolic change takes place in one or both cells such that A's efficiency, as one of the cells firing B, is increased.' D. O. Hebb, *The Organization of Behavior* (Wiley & Sons, New York, 1949). The phrase 'neurons wire together if they fire together' was stated by Siegrid Löwel and Wolf Singer in their paper: S. Löwel and W. Singer, *Science* (1992) 255, 209–212.

4   You can try this gorilla illusion for yourself here: https://tinyurl.com/pej6jcl. Even if you have seen this before, have a go. This is a new version with a twist!

5   Not only are new experiences evaluated in terms of our existing neural networks, these experiences then modify and update our neuronal connections further. So our minds are really changing every second. A couple of nice books on this are: S. Greenfield, *Mind Change* (Penguin Random House, London, 2014); and S. Seung, *Connectome*.

6   Of course, there is the question as to whether you really 'see' in your mind's eye the same pink or green colour that I do. It is impossible to verify this, but the relational difference between the colours and the objects in the world which they refer to is consistent between us, so that although we cannot be completely sure they are the *same* concepts, they are at the very least, equivalent.

7   Donald and Stuart Geman's assertion is that the major advances
    in human knowledge were all made decades ago, with relatively
    few new notable great discoveries in the last 50 years. This in itself
    is somewhat questionable, let alone the argument that enhanced
    digital connectivity between humans is responsible for the
    decline. D. Geman and S. Geman, *PNAS* (2016), 113, 9384–9387.
8   These examples of multiple different inventors of the same
    technology are given by Kevin Kelly. The writer Matt Ridley
    describes how the invention of the incandescent bulb, or some
    very close version of it, can be attributed to no fewer than 23
    people, though Thomas Edison receives all the credit in popular
    history, while Alexander Graham Bell is well known for inventing
    the telephone, yet another inventor, Elisha Gray, filed for a patent
    on the telephone on the very same day as Bell. M. Ridley, *The
    Evolution of Everything* (Fourth Estate, London, 2015); K. Kevin,
    What Technology Wants (Penguin, London, 2011).
9   P. L. Jackson et al., *NeuroImage* (2005) 24, 771–779.
10 S. Seung, *Connectome*.
11 L. M. Mujica-Parodi, *PLOS ONE* (2009) 4, e6415.
12 P. B. Singh et al., *Chemical Senses* (2018) 43, 411–417.

Chapter 7

1   The details of this harrowing episode in the small community's
    history are sourced from C. Wilkie et al., *Can J Psychiatry* (1998)
    43, 823–828.
2   P. Hedström et al., *Social Forces* (2008) 87, 713–740.
3   Cited in 'World Wide Warp', *New Scientist*, February 2016.
4   Y. N. Harari, *Sapiens* (Vintage Books, London, 2011).
5   I. Glynn and J. Glynn, *The Life and Death of Smallpox* (Profile
    Books, London, 2005).
6   A. L. Schmidt et al., *Vaccine* (2018) 36, 3606–3612.
7   Data from World Health Organisation (WHO). Global cases
    from News Bulletin 29 November 2018: 'Measles cases spike

globally due to gaps in vaccination coverage'. European Region data from 'Measles cases and MCV1 & MCV2 coverage in the WHO European Region, 2009–2018'.

8   For more examples, see the work of social scientists Daniel Lieberman and Emily Falk who have investigated cases of ideas and opinions leaping between people 'across the blood brain barrier' and spreading rapidly through social networks.

9   https://www.patientslikeme.com/.

10 http://crisismappers.net/.

11 http://ejatlas.org/.

Chapter 8

1   M. L. King Jr., 'Interconnected World' sermon, Ebenezer Baptist Church, Atlanta, USA (1967).

2   The full essay by Leonard Read, written in 1958 and published in the December 1958 issue of *The Freeman* can be found at http://www.econlib.org/library/Essays/rdPncl1.html#firstpage-bar (acknowledgements to The Foundation for Economic Education, Inc.).

3   The economist Adam Smith first coined this phrase.

4   J. Kleinberg, '*E pluribus unum*' in J. Brockman (ed.), *This Will Make You Smarter* (HarperCollins, New York, 2012), p. 74.

Chapter 9

1   This figure of 64,285,009 km of roads relates to the year 2013 from the CIA World Factbook (https://www.cia.gov/library/publications/the-world-factbook/). It is obviously quite a large figure: those roads would run to the moon and back almost 84 times.

2   This total figure for number of vehicles relates to the year 2012 and comprises 808.7 million registered cars and 291 million registered commercial vehicles (https://www.statista.com/statistics/281134/number-of-vehicles-in-use-worldwide/). The

number of ships is for 2016 and comprises 16,892 bulk carriers, 10,919 cargo ships, 7,065 crude oil tankers, 5,239 container ships, 5,204 chemical tankers, 4,316 passenger ships and 1,770 liquefied natural gas tankers (https://www.statista.com/statistics/264024/number-of-merchant-ships-worldwide-by-type/).

3 To see a quite startling real time display of all the passenger planes currently in flight check out https://www.flightradar24.com. This satellite tracking can apparently miss some planes but it reports between 6,000 and 10,000 planes in the air at any one time. For example, checking at 7:21 on 15.02.17 there were 9,122 planes in the air. If each plane carried 200 people on average (and it can range from over 500 for Airbus 380 to just several people for a private jet) then that equates to over 1.8 million people being suspended thousands of metres in the air at that point in time! For further interest, an impressive simulation video showing the complex flight patterns over the UK created by air traffic management company NATS can be seen at http://www.nats.aero/news/take-guided-tour-around-uk-airspace/. The figure for number of airports worldwide comes from the CIA World Factbook (https://www.cia.gov/library/publications/the-world-factbook/). There were 41,820 airports in 2016 (the top three most visited by passengers being Atlanta, Beijing and Dubai).

4 Friends of the Earth Europe, Sustainable Research Institute and GLOBAL 2000 'Overconsumption? Our use of the world's natural resources', https://www.friendsoftheearth.co.uk/sites/default/files/downloads/overconsumption.pdf.

5 Arthur Tansley (1871–1955) was an English botanist and pioneer of ecology.

6 You can read about this rare conservation success story in the excellent summary paper by ecologist Jeremy Thomas; J.A. Thomas, *Science* (2009) 325, 80–83.

7 Mora et al., *PLoS Biology* (2011) 9, e1001127.

8 This appropriation by humans of the materials produced by

plants and other autotrophic organisms at the bottom level of the biomass pyramid is measured by the HANPP (human appropriation of net primary production) index. Large HANPP values mean there is far less material and energy available to support wild species populations. F. Krausmann et al., *PNAS* (2013) 110, 10324–10329.

9   'The battle for the soul of biodiversity', *Nature* (2018) 560, 423–425; https://www.nature.com/articles/d41586-018-05984-3.

10 As of May 2019, when the first IPBES global assessment report was published, there were 132 UN member states involved, with 455 scientists from 50 countries authoring the 1,800-page report.

11 Diaz et al., *Science* (2018) 359, 270–272.

12 Mora et al., *PLoS Biology* (2011) 9, e1001127.

13 C. Darwin, *On the Origin of the Species* (1859).

Part 3

1   This poem, 'A Reflection at Sea', appears in the anthology *The Poetical Works of Thomas Moore* (1841). I first encountered the poem on an old embroidery piece at Basildon House, Berkshire with the signature underneath: 'Hannah Dracup'.

Chapter 10

1   Brian Eno, 'Ecology' in Brockman (ed.), *This Will Make You Smarter*, p. 294.

2   For example, in the USA, normally a scientific pioneer but in the muddy backwaters in regards to public acceptance of evolution, a Pew Research Centre report in 2009 reported: 'A majority of the public (61 per cent) says that human and other living things have evolved over time, though when probed only about a third (32 per cent) say this evolution is "due to natural processes such as natural selection" while 22 per cent say "a supreme being guided the evolution of living things for the purpose of creating humans and other life in the form it exists today."

Another 31 per cent reject evolution and say that "humans and other living things have existed in their present form since the beginning of time." http://www.people-press.org/2009/07/09/section-5-evolution-climate-change-and-other-issues/.

3 These examples are cited in the Bruce Hood's excellent book *The Self Illusion* (Constable, London, 2011). The first experiment is from a pioneer in human psychology Elizabeth Loftus (E.F. Loftus, *Cognitive Psychology* (1975) 7, 560–72). The second is from: K. A. Wade et al., *Psychonomic Bulletin & Review* (2002) 9, 597–603.

4 A cognitive bias is a systematic pattern of deviation from rational judgment. A number of types of bias have been identified and their consistency across humans from different cultures suggests they are probably adaptive. An excellent book, *Thinking, Fast and Slow* by Daniel Kahneman, the Nobel Prize winner for Economics, is a great introduction to these (Penguin Books, London, 2012).

5 A. R. Wood, *American Journal of Sociology* (1930) 5, 707–717.

6 C. H. Cooley, *Human Nature and the Social Order* (Charles Scribner's Sons, New York, 1902). https://archive.org/details/humannaturesocia00cooluoft/.

7 S. Greenfield, *Mind Change: How 21st century technology is leaving its mark on the brain* (Penguin Random House, London, 2014).

8 D. C. Dennett, *Consciousness Explained* (Little, Brown, Boston, 1991).

9 M. D. Lieberman, *Social: Why our brains are wired to connect* (Oxford University Press, Oxford, 2013).

10 T. Metzinger, 'Cognitive Agency' in Brockman (ed.), *This Idea Must Die* (Harper Perennial, 2015), p. 150. See also interview with Michael Taft: https://deconstructingyourself.com/what-is-the-self-metzinger.html.

11 M. Pagel, *Wired For Culture* (Penguin Books, London, 2012).

12 Z. Cormier, *Nature News* (2016) doi:10.1038/nature.2016.19727.

13 Each survey had over 1,000 participants who were regular meditators (practising at least once a week). Schlosser et al., *PLoS ONE* (2019) 14, e0216643; C. Vieten et al., *PLoS ONE* (2018) 13, e0205740.

Chapter 11

1 An excellent collation of this evidence of broad differences in systems of thought between cultures is Richard E. Nisbett's *The Geography of Thought* (Nicholas Brealey, London, 2003).

2 M. Haig, *Reasons to Stay Alive* (Canongate, Edinburgh, 2015).

3 The summary of the moral education promoted by Confucius presented here is derived from I. P. McGreal (ed.), *Great Thinkers of the Eastern World*. The quote by Henry Rosemont is included in Richard Nisbett's *The Geography of Thought*.

4 This study focuses on two 'prototypically individualist cultures', the United States and Germany, and two 'collectivist' cultures, Russia and Malaysia. U. Kühnen et al., *Journal of Cross-Cultural Psychology* (2001) 32, 366–372. The study on autistic individuals is from R. A. Almeida et al., *Neuropsychologia* (2010) 48, 374–381.

5 This experiment, conducted by developmental psychologists, involved asking four- to six-year-old children to report on daily events, during which the psychologists recorded the number of self-references. J. J. Han et al., *Developmental Psychology* (1998) 34, 701–713.

6 R. E. Nisbett, *The Geography of Thought*.

7 William James, with this term 'blooming, buzzing confusion' from his 1890 book *Principles of Psychology*, describes how a baby might first experience the world before it has learned to separate the inputs from all its senses ('assailed by eyes, ears, nose, skin, and entrails at once') and categorise these into distinct 'objects'. Similar ideas were followed by seventeenth-century philosopher John Locke, who described babies as blank slates ('tabula rasa')

that are inscribed upon with experience. Of course, experience is essential in developing our internal reference library (the abstract concepts) of objects that exist in the world. However, recent work has also shown that babies are born with strong predispositions and discriminate between certain things better than others. For example, they are able to distinguish between human faces better than faces of other species, as well as being 'programmed' to learn languages rapidly. So the idea of a blank slate is perhaps not completely correct, but James is no doubt right that a baby emerging into the world, before they have learnt to order objects neatly in their heads, will still experience it as a pretty wild and confusing place!

8 H. C. Santos et al., *Psychological Science* (2017) 28, 1228–1239.
9 E. Diener and S. Oishi, *Psychological Enquiry* (2005) 16, 162–167.
10 Eisenberger et al., *Science* (2003) 302, 290–292.
11 S. Pinker, *The Village Effect* (Atlantic Books, London, 2014).
12 Examples here are Richard Louv, who in his book *Last Child in the Woods* (Workman Publishing New York, 2005) introduced the term 'nature deficit disorder'. The suggestion that disengagement with nature is a recent phenomenon is questioned by others such as Elizabeth Dickinson who propose that the problem is much older. E. Dickinson, *Environmental Communication* (2013) 7, 315–335.
13 M. Richardson et al., 'The Impact of Children's Connection to Nature' (2015), a report for the Royal Society for the Protection of Birds (RSPB), http://ww2.rspb.org.uk/Images/impact_of_children's_connection_to_nature_tcm9-414472.pdf.
14 RSPB 'Connecting with Nature: finding out how connected to nature the UK's children are' (2013), http://ww2.rspb.org.uk/Images/connecting-with-nature_tcm9-354603.pdf.
15 N. Christakis, 'Holism' in *This Will Make You Smarter*; F. Vester, *The Art of Interconnected Thinking* (Mcb Verlag, Munich, 2012).
16 F. Vester, *The Art of Interconnected Thinking*.

17 The Necker cube optical illusion is named after Swiss
    crystallographer Louis Albert Necker who conceived it in 1832.

Chapter 12

1   The date of the Buddha's lifetime is disputed across many
    different texts, but from a German conference on the topic the
    most accepted date among scholars is around 400 BC. L. S.
    Cousins, *Journal of the Royal Asiatic Society* (1996) 6, 57–63.
2   I. P. McGreal (ed.), *Great Thinkers of the Eastern World*
    (HarperCollins, New York, 1995), p. 163.
3   Y. N. Harari, *Sapiens*.
4   In *Civilisation and its Discontents* (1930), Freud writes: 'The ego
    seems to maintain clear and sharp lines of demarcation. There is
    only one state – admittedly an unusual state, but not one that can
    be stigmatised as pathological – in which it does not do this. At
    the height of being in love, the boundary between ego and object
    threatens to melt away. Against all the evidence of the senses,
    a man who is in love declares that "I" and "you" are one, and is
    prepared to believe it were a fact.'
5   Newspaper columnist and writer George Monbiot describes
    how our language is changing to reflect the increasing tendency
    towards competitive individualism: 'Our most cutting insult
    is "loser". We no longer talk about people. Now we call them
    "individuals". We prefix many terms needlessly with "personal"
    such as "my personal belongings", "my personal preference". The
    language all around us seems to reinforce this sense of isolated
    autonomy.' G. Monbiot, *How Did We Get into This Mess?* (Verso,
    London, 2014).
6   M. Pagel, *Wired For Culture*.
7   J. C. Flack and F. B. M. de Waal, *Journal of Consciousness Studies*
    (2000) 7, 1–29.
8   Note, this is not to state there is no natural selection operating in
    modern human populations; there are some clear demonstrations,

for example, evolution of resistance against diseases experienced at a pre-reproductive age. Rather, it simply states that the impact of natural selection in modern societies is much less than in previous historic times because of our social support systems (hospitals, GPs, mental health support, medical inventions such as prosthetic limbs, sight-correcting glasses and contact lenses), which all buffer us from increased mortality risks.

9   Mortality rates from road traffic accidents range from 3 to 74 per 100,000, depending on which country you live in. The lowest rates are for Denmark and the Maldives, the highest for Libya (http://www.who.int/violence_injury_prevention/road_safety_ status/2015/TableA2.pdf ?ua=1).

10 These empirical examples of cheats across different animal groups are cited in C. Riehl and M. E. Frederickson, *Philosophical Transactions of the Royal Society B: Biological Sciences* (2016) 371. Theoretical exploration as to why cheats prosper in larger groups can be found in R. Boyd and P. J. Richerson, *Journal of Theoretical Biology* (1988) 132, 337–356.

11 There is little sign yet of the global obesity crisis reversing, with obesity doubling since 1980, and with 13 per cent of the world's population classified as obese in 2014 (39 per cent overweight; http://www.who.int/news-room/fact-sheets/detail/obesity- and-overweight). However, in some countries, such as the UK (where the problem is already more severe with a quarter of the population obese), there is at least some evidence that rates of increase are slowing, if not yet reversing (http://www.nhs.uk/ news/2013/09September/Pages/Obesity-in-England-rising-at-a- slower-rate.aspx).

Chapter 13

1   Figures here are from references within the Wikipedia entry for kodokushi https://en.wikipedia.org/wiki/Kodokushi accessed 16.4.2017.

2 http://www.japantimes.co.jp/
   news/2011/10/09/national/media-national/
   nonprofits-in-japan-help-shut-ins-get-out-into-the-open.

3 UK 2014 loneliness poll http://opinium.co.uk/lonely-and-
   starved-of-social-interaction/ and http://opinium.co.uk/
   busy-lives-but-lonely-britain.

4 The poll was the European Quality of Life Survey, 2011/12,
   and is reported in the UK ONS survey: http://webarchive.
   nationalarchives.gov.uk/20160107113746/ http://www.ons.gov.
   uk/ons/dcp171766_393380.pdf.

5 G. Monbiot, *How Did We Get into This Mess?*

6 This is especially true in cities. Those living in areas with a
   dense population such as inner cities are more likely to describe
   themselves as lonely (24 per cent) compared to villages (13 per
   cent) http://opinium.co.uk/lonely-britain.

7 http://www.telegraph.co.uk/news/politics/10909524/Britain-
   the-loneliness-capital-of-Europe.html.

8 http://opinium.co.uk/busy-lives-but-lonely-britain.

9 As mentioned in Chapter 11, the phenomenon of lack of
   connectedness to nature and consequent health impacts has
   been dubbed 'nature deficit disorder'. (R. Louv, *Last Child in the
   Woods*) Since the term was first coined there has been extensive
   research into this area, e.g. R. Lovell et al., *Journal of Toxicology
   and Environmental Health, Part B* (2014) 17, 1–20; D. Cox et
   al., *International Journal of Environmental Research and Public
   Health* (2017) 14, 172.

10 This description comes from Ian McGreal's *Great Thinkers of the
   Eastern World*, which describes Hindu lore from the Upanishads
   explaining how humans often identify with the bodily self
   rather than the true self, termed *Atman* (spirit) which is actually
   equivalent to *Brahman*, the Universal Spirit.

11 Galatians 3:28.

12 The ultimate example of this view of the world as a set of abstract
   discrete entities is Plato's Theory of Forms, which argues that

non-physical ideas represent the most accurate reality. The thesis is that every object in the world is a manifestation of an abstract unchanging 'blueprint'. For example, all horses are a manifestation of the perfect original idea of a Horse, which has existed eternally. Reflecting on the discovery of Evolution (which to be fair, Plato had no idea about), we can see that this doesn't make sense. Animals evolve, not towards some predetermined form, but in a haphazard fashion depending on the best solutions to recent environmental conditions. Before horses evolved from small, dog-sized equid ungulates, there were no such thing as a horses, real or abstract. As an aside, however, there is a slight conundrum here. In the human mind, there exists the potential for any thought that humans will ever have, through the possible connections between billions of neurons, even if only a very small subset of these connections are realised in any given person. So somewhere in all of our brains is the *potential* for an abstract conception of any animal or object that ever existed, or *ever will exist*. So maybe Plato's idea is not quite as wrong as first thought.

13 Thich Nhat Hanh, 'Healing Pain and Dressing Wounds', in B. Boyce (ed.), *In the Face of Fear: Buddhist Wisdom for Challenging Times* (Melvin Mcleod, 2009).

14 This text from Ishopanishad: sloka 6, 7 comes from the Advaita Vedanta school of Hindu philosophy, which emphasises non-dualism: that the universe is one essential reality, and that all facets and aspects of the universe is ultimately an expression or appearance of that one reality. Orlando O. Espín and James B. Nickoloff, *An Introductory Dictionary of Theology and Religious Studies* (Liturgical Press, 2007).

15 R. E. Nisbett, *The Geography of Thought*.

16 J. Nakamura, M. Csíkszentmihályi, 'Flow Theory and Research' in C. R. Snyder, Erik Wright and Shane J. Lopez, *Handbook of Positive Psychology* (Oxford University Press, 2001), pp. 195–206.

Chapter 14

1 The transcript of this letter can be found at http://murderpedia. org/male.W/images/whitman_charles/docs/typewritten_letter. pdf along with other letters, such as that of the heroic Co-op supervisor who volunteered to help the police and was given a rifle and climbed the tower with them to finally kill the gunman.

2 This visit took place 29 March 1966. The doctor gave no medication but recommended that Whitman come back a week later and, if he needed to see a therapist in the meantime, he could let him know. Whitman did not return and went on to murder sixteen people in August 1966. Source: http:// murderpedia.org/male.W/images/whitman_charles/docs/ heatley.pdf

3 https://www.theatlantic.com/magazine/archive/2011/07/ the-brain-on-trial/308520.

4 B. Hood, *The Self Illusion*.

5 M. Rutter and T. G. O'Connor, *Developmental Psychology* (2004) 40, 81.

6 https://www.theatlantic.com/magazine/archive/2011/07/ the-brain-on-trial/308520

7 A. Caspi et al., *Science* (2002) 297, 851–854.

8 B. Hood, *The Self Illusion*.

9 B. Spinoza, *Ethics* (1677).

10 J. Tooby, 'Nexus Causality, Moral Warfare and Misattribution Abritrage' in *This Will Make You Smarter*, pp 35–37.

11 This is a classic cognitive bias – when bad things happen to us we are more likely to attribute them to outside factors beyond our control, while when good things happens to us we attribute them to our innate skills or personality. Clearly, this protects and reinforces our self-esteem. In contrast, for others, it is often the exact opposite: when bad things happen to them, or are caused by them, we blame the characteristics of the individual themselves,

yet, when good things happen to them, we explain this away as 'luck' from a favourable set of external circumstances (allowing us to protect our perceived relative standing compared to the other person).

12 M. W. Morris and K. Peng, *Journal of Personality and Social Psychology* (1994) 67, 949.

13 I. Choi, R. Dalal, C. Kim-Prieto and H. Park, *Journal of Personality and Social Psychology* (2003) 84, 46.

14 The DNA is affected through a process of 'methylation', where methyl groups are added to the DNA molecule, which changes its functioning even without changing the sequence of nucleotide base pairs. Studies in mice confirm this link between adult experience (e.g. exposure to chemical pollutants) affecting offspring obesity through DNA methylation, but a recent synthesis of research in humans suggests the jury is still out on the importance of epigenetic effects in driving obesity: S.J. van Dijk et al., *Int J Obes* (2015) 39, 85–97.

15 Nicholas A. Christakis and James H. Fowler, *Connected: The surprising power of our social networks and how they shape our lives* (Little, Brown, New York, 2009).

Part 4

1 I wrote this short set of words as a meditation tool. They broadly capture the sentiment of this book. I include them here in the hope that others might find them useful too.

Chapter 15

1 J. Hillman, 'A Psyche the size of the earth: a psychological foreword' in T. Roszak, M. E. Gomes and A. D. Kanner (eds), *Ecopsychology- Restoring the Earth, Healing the Mind* (Counterpoint, Berkeley, 1995).

2 Amy Proal's blog can be found at http://microbeminded.com and one of her papers investigating cause of chronic fatigue

syndrome is A. D. Proal et al., *Immunologic Research* (2013) 56, 398–412.

3   For reviews of microbiome links to neurological conditions see either Sampson et al., *Cell Host and Microbe* (2015) 17, 565–576, or H. Tremlett et al., *Ann Neurol.* (2017) 81, 369–382.

4   European Environment Agency (2018). Chemicals for a sustainable future. EEA Report No 2/2018. doi: 10.2800/92493.

5   Miles Richardson writes a blog about nature connectedness that can be found at https://findingnature.org.uk.

6   UK Government Department for Environment and Rural Affairs, 'Evidence Statement on the Links Between Natural Environments and Human Health', May 2017.

7   The company Velux have produced a number of reports from independent research on how much time we spend indoors and how we can potentially be exposed to pollutants while lacking natural daylight ('*The Indoor Generation*' https://www.velux. com/article/2018/indoor-generation-facts-and-figures). Other independent studies also confirm this figure of spending 90 per cent of time indoors. For example, a 2001 study carried out by the US Environmental Protection Agency (The National Human Activity Pattern Survey) found Americans spend a worrying 87 per cent of time indoors and 6 per cent in vehicles.

8   G. Engel, *Science* (1977) 196, 129–136.

9   Source for America: National Center for Health Statistics, 'Antidepressant Use Among Persons Aged 12 and Over: United States, 2011–2014', NCHS Data Brief No. 283, August 2017. Source for England: NHS Business Services Authority, 'Antidepressant prescribing 2015/16 and 2016/17', https:// www.nhsbsa.nhs.uk/prescription-data/prescribing-data/ antidepressant-prescribing.

10 The biologist Stephen J. Gould proposed that science has the appropriate tools to teach about the factual character of the natural world, while religion operates in the realm of human

purposes, meanings, and values – subjects that the factual domain of science might illuminate, but can never resolve. S. J. Gould, *Rocks of Ages: Science and Religion in the Fullness of Life* (Ballantine Books, New York, 1999).

11 H. C. Santos et al., *Psychological Science* (2017) 28, 1228–1239.

12 The study showing increases in narcissism in college students since the 1980s controls for variance across campuses and ethnic groups (J. M. Twenge and J. D. Foster, *Social Psychological and Personality Science* (2010) 1, 99–106). Levels of narcissism also increased across all ethnic groups tested (J. M. Twenge and J. D. Foster, *Journal of Research in Personality* (2008) 42, 1619–1622). If you want to take the test yourself you can find a version at https://openpsychometrics.org/tests/NPI/1.php.

13 The clinical definition of the more severe Narcissism Personality Disorder can be found in the DSM-IV and DSM-5 Criteria for the Personality Disorders. The argument that NPD has increased over time is made by Twenge and Campbell in their 2009 book *The Narcissism Epidemic*. From a National Institute of Health study, 35,000 people were asked to recall symptoms in their lives that researchers identified as NPD. The argument goes that if rates of NPD were constant over time then *more* older people should report experiencing symptoms at some point in their lives. In contrast, 1 in 10 young Americans in their twenties had experienced symptoms, compared to only 1 in 30 of those over sixty-five. One explanation is that older people forgot experiencing these symptoms over the course of their lives, but Twenge and Foster think this is unlikely. Instead, they suggest we may be experiencing an epidemic in narcissism, which is worsening over time.

14 J. M. Twenge et al., *Journal of Personality and Social Psychology* (2012) 102, 1045–1062.

15 Bhutan has created a Gross National Happiness Index to gauge its progress rather than the standard focus of most countries on economic metrics such as gross domestic product (a monetary

measure of all the goods and service produced each year). The four pillars of Bhutan's Gross National Happiness are: 1) sustainable and equitable socio-economic development; 2) environmental conservation; 3) preservation and promotion of culture; and 4) good governance. K. Ura et al. (2012), *An Extensive Analysis of GNH Index* Thimphu, Bhutan: The Centre for Bhutan Studies.

16 W. K. Campbell et al., *Personality and Social Psychology Bulletin* (2005) 31, 1358–1368.

17 It is not just celebrity magazines in which the culture of narcissism and values of extreme individuality are propagated. A large number of studies have looked at how the use of individualistic words and phrases are now much more common in our mainstream newspapers. H. E. Nafstad et al., *Journal of Community & Applied Social Psychology* (2007) 17, 313–327; J. M. Twenge et al., *Journal of Cross-Cultural Psychology* (2013) 44, 406–415.

18 The video game which improves empathy can be read about here reviewed at https://centerhealthyminds.org/news/ video-game-changes-the-brain-and-may-improve-empathy-in-middle-school-children, and reported in the paper by T. R. A. Kral et al., *npj Science of Learning* (2018) 3, 13.

19 You can watch the empathy staring test organised by Derren Brown: https://www.youtube.com/watch?v=tEnYAUvlTS8. Note, the DNA test prior to this staring test may also have helped the participant in entering the strong state of empathy by breaking down deeply held beliefs that he was intrinsically different to people from other countries. This approach has been shown in a moving video by the firm Momondo that went viral on the internet, being viewed over 18 million times. People with racist views were filmed as results from DNA tests were revealed (https://www.youtube.com/watch?v=tyaEQEmt5ls).

Chapter 16

1   A. J. Montiel-Castro et al., *Frontiers in Integrative Neuroscience* (2013) 7.

2   This is thought to operate through bacteria in our gut producing products that influence enterendocrine cells in the gut lining. These cells were first thought to link to the central nervous system (interacting with the vagus nerve cells surrounding the gut) exclusively through hormones (see E. A. Mayer, *Nature Reviews Neuroscience* (2011) 12, 453), but more recently it has been revealed that they can communicate rapidly using the neurotransmitter chemical serotonin (see B. U. Hoffman and E. A. Lumpkin, *Science* (2018) 361, 1203–1204).

3   E. A. Mayer, *Nature reviews Neuroscience* (2011) 12.10.1038/ nrn3071.

4   L. Van Oudenhove et al., *The Journal of Clinical Investigation* (2011) 121, 3094–3099.

5   To learn more about the effects of foods on our gut microbiome and in turn how this can affect our mental states see: T. G. Dinan et al., *Biological Psychiatry* (2013) 74, 720–726 and H. Wang et al., *J Neurogastroenterol Motil* (2016) 22, 589–605.

6   These gut-to-brain pathways may open up opportunities for pharmaceutical or nutritional approaches to treat autism spectrum disorders. For some example of diet-based treatments see: C. G. de Theije et al., *Eur. J. Pharmacol* (2011) 668, S70–S80.

7   C. Frantz et al., *Journal of Environmental Psychology* (2005) 25, 427–436.

8   In both rats and monkeys changes in the microbiome result from stress caused by temporary maternal separation, although this only seems to be linked to altered cortisol levels in rats. M. T. Bailey and C. L. Coe, *Dev. Psychobiol.* (1999) 35, 146–155; S. M. O'Mahony et al., *Biological Psychiatry* (2009) 65, 263–267.

9   https://www.humi.site. See also G. A. Rook, *PNAS* (2013) 110, 18360–18367.

10 M. H. Davis et al., *Journal of Personality and Social Psychology* (1996) 70, 713.

11 A. Aron et al., *Journal of Personality and Social Psychology* (1991) 60, 241–253; A. Aron et al., *Journal of Personality and Social Psychology* (1992) 63, 596–612.

12 For more about this dichotomy and the gradual integration between the two research fields see E. A. Bragg, *Journal of Environmental Psychology* (1996) 16, 93–108.

13 The hypothesis of three clusters of values related to self-identity (egoic, social-altruistic, and biospheric) is outlined in P. C. Stern and T. Dietz, *Journal of Social Issues* (1994) 50, 65–84. For the rejection of the distinction between the social-altruistic and biospheric clusters, and the suggestion of a generalised 'self-transcendent' cluster, see W. P. Schultz, *Journal of Environmental Psychology* (2001) 21, 327–339. Other research has shown how different people develop at different rates and extents towards this self-transcendence. In one perspective only some adults achieve a stage of 'Self Authoring' associated with a generally well-adjusted adulthood and even fewer more to a stage of 'Self-Transforming' associated with heightened empathy and the ability to hold differing perspectives. For more background see A. H. Pffanenberger, P. W. Marko, A. Combs (eds), *The Postconventional Personality* (State University of New York Press, 2001).

14 A. L. Metz, *Vistas Online* (2017) 11, 1–14.

Chapter 17

1 E. A. Caspar et al., *Current Biology* (2016) 26, 585–592.

2 J. L. Greene, T. Cowan, 'Table Egg Production and Hen Welfare: Agreement and Legislative Proposals', CRS Report for Congress (2014), 42534. https://fas.org/sgp/crs/misc/R42534.pdf.

3 Thousands of years ago our global supply chains were mesmerisingly complex. As historian Peter Frankopan describes:

'Two millennia ago, silks made by hand in China were being worn by the rich and powerful in Carthage and other cities in the Mediterranean, while pottery manufactured in southern France could be found in England and in the Persian Gulf. Spices and condiments grown in India were being used in the kitchens of Xingjiang, as they were in those of Rome. Buildings in northern Afghanistan carried inscriptions in Greek, while horses from Central Asia were being ridden proudly thousands of miles away to the East.' If that was the case thousands of years ago, think of the twenty-first-century network of global trade that make the impacts of our consumer decisions immediately global and so dilute that it is hard to delineate any clear moral responsibility.

4   According to UN figures, in 2014, 54 per cent of the world's population lived in urban areas, with the figure expected to be 66 per cent by 2050: http://www.un.org/en/development/desa/news/population/world-urbanization-prospects-2014.html.

5   Steve Hilton, *More Human* (WH Allen, London, 2015).

6   The term 'apophenia' was coined by German neurologist Klaus Conrad to describe a form of mental illness, but, according to psychologist David Pizarro, many normal people suffer from the condition to some degree in their tendency to develop superstitions based on spurious correlations. Other studies show that babies have an innate tendency to develop such spurious connections (Hood, 2008). D. Pizarro, 'Everyday Apophenia' in *This Will Make You Smarter*; B. Hood, *Supersense* (Constable, London, 2009).

7   M. McCarthy, *The Moth Snowstorm: Nature and Joy* (John Murray, London, 2015).

8   As the biochemist and systems thinker Frederic Vester describes the failure of governance: 'Under pressure from short-term necessities, our political and economic decision-makers show little sign of acknowledging interconnectedness – let alone taking it into account in their plans and actions ... The accumulated errors of recent years have shown beyond doubt that the classic

approaches to planning, whether in business, in regional planning, in development aid, or in environmental policy, have all failed (indeed, could not but fail) because of the increasingly complex network of effects and repercussions that they leave out of account.' Frederic Vester, *The Art of Interconnected Thinking*.

9   Arne Naess, *Ecology of Wisdom* (Penguin Random House, London, 2008). Naess's theories about the realisation of a more interconnected 'ecological self' which transcends the ego have similarities with other philosophies such as Mahayana Buddhism. The writer Alan Drengson in his piece 'Ecophilosophy, Ecosophy and the Deep Ecology Movement: An Overview' describes how the extension of self and the idea of the ecological Self overlaps in many ways with work in transpersonal psychology/ecology with parallels in Mahayana Buddhism as well as certain minority Christian philosophies (http://www.ecospherics.net/pages/ DrengEcophil.html).

10 Other benefits from helping others in our close social 'in-groups' include the possibility that we might increase our standing in the group or increase our group's overall success against competing groups. In addition to the increasing the chance of obtaining reciprocal aid from others and the benefiting of our kin, these factors likely led to the evolution of helping behaviour in in-groups.

11 Summarised in T. Roszak, M. E. Gomes and A. D. Kanner (eds), *EcoPsychology – Restoring the Earth, Healing the Mind*.

12 M. J. Zylstra et al., *Springer Science Reviews* (2014) 2, 119–143.

13 E. Gosling and K. J. H. Williams, *Journal of Environmental Psychology* (2010) 30, 298–304.

14 In the face of the economic system taking into account the 'externalised' costs on the natural world, many modern conservationists now advocate a 'natural capital approach' where the values of nature are factored into economic decision making. This seeks to protect nature through economic means, but completely fails to tackle our consumer behaviours that

are the root cause of the pressures. In fact, some authors warn that such a materialistic approach might backfire by further promoting consumerist mindsets (e.g. see the 2009 WWF report by T. Crompton and T. Kasser, *Meeting Environmental Challenges: The role of human identity* (WWF-UK, Surrey, ISBN: 9781900322645).

15 Collective human perspectives about the world determine our institutions, but we should also recognise that there are further feedbacks where our institutions can, in turn, shape our perceptions. Author Kate Raworth in her book *Doughnut Economics* bemoans the facts that currently such feedback processes seem to be driving a culture for both humans and their cultures to become more individualistic. However, on an optimistic note she points out that we can reverse this process, and it starts with reversing our fundamental beliefs about our relationships with the world around us. K. Raworth, *Doughnut Economics: Seven Ways to Think Like a 21st-Century Economist* (Random House Business, London, 2017).

16 WWF Living Planet Report (2016): http://awsassets.panda.org/ downloads/lpr_living_planet_report_2016.pdf.

Chapter 18

1 Eurostat per capita figures for 2017 across the EU28 countries are: 13.6 tonnes of raw material consumed per year and production of greenhouse emissions equivalent to 8.7 tonnes of CO_2. Energy use in terms of kg oil equivalent are for the European Union in 2015 sourced from the World Bank (3,207 kg/person).

2 This calculation is on the basis that modern Homo sapiens is thought to have evolved between 350,000 to 260,000 years ago (Schlebusch et al. 2017). Yet only over the last 60 years, dubbed the great acceleration, have we seen near exponential increase in many consumption related variables such as energy use, water use, fertiliser consumption, transportation that have led to huge

impacts on our natural world. W. Steffen et al., *The Anthropocene Review* (2015) 2, 81–98). C. M. Schlebusch et al., *Science* (2017) 358, 652–655.

3   In 2016, the role of social media, including fake news and bots, was implicated in influencing voting in the US elections and the UK referendum to leave the European Union.

4   For more information, see D. Goleman and R. L. Davidson, *The Science of Meditation* (Penguin Random House, London, 2017).

5   The reduced intergroup bias experiment described is Y. Kang et al., *Journal of Experimental Psychology: General* (2014) 143, 1306–1313. Subsequent meta-analysis of such approaches supports these results: X. Zeng et al., *Frontiers in Psychology* (2015) 6.

6   An example here is the Kindness Curriculum for helping to develop empathy traits in young children developed by the Center for Healthy Minds at the University of Wisconsin-Madison.

7   P. W. Schultz, *Journal of Social Issues* (2000) 56, 391–406.

8   Social media is increasingly linked to self-rumination and mental health problems. For example, a recent report by the UK National Health Service found that almost a quarter of eleven- to sixteen-year-old girls have a mental health disorder and a third of these have attempted self-harm or suicide. These children are more likely to spend long periods on social media comparing themselves with others while admitting that it strongly affects their mood. *Mental Health of Children and Young People in England*, 2017, NHS digital.

9   P. M. Gollwitzer, *American Psychologist* (1999) 54, 493–503.

10   H. E. Nafstad et al., *Journal of Community & Applied Social Psychology* (2007) 17, 313–327; J. M. Twenge et al., *Journal of Cross-Cultural Psychology* (2013) 44, 406–415.

11   The Marxist philosopher Antonio Gramsci argues that the dominant political power doesn't just wield military and economic power, it also makes ideas seem like common sense

through language. He suggests this battle for ideas may explain why left-wing political movements fail to gain significant traction in the twentieth century. The same may be true for the stalling of environmental movements. Some researchers from the WWF (T. Crompton and T. Kasser, *Meeting Environmental Challenges: The role of human identity*) have suggested that a failure to develop new values but to adopt materialistic paradigm (e.g. using terms such as 'natural capital' and 'ecosystem services') could be a grave error. They describe how 'the willingness with which mainstream environmentalism has embraced self-enhancing, materialistic values and life goals has actually served to reinforce the dominance of these values and goals'. Such values are associated with more negative environmental attitudes and damaging environmental behaviour. See also T. Jackson, *Motivating Sustainable Consumption: A Review of Evidence on Consumer Behaviour and Behavioural Change*, Report to the Sustainable Development Research Network (2004), Centre for Environmental Strategy, University of Surrey.

12 T. Kasser and A. D. Kanner (eds), *Psychology and Consumer Struggle: The struggle for a good life in a materialistic world* (American Psychological Association, Washington D.C., 2004).

13 The exceptions are a handful of countries like China where there are extensive attempts to reduce individualistic content in the media and on the internet, combined with a far-reaching social modelling programme aiming to increase group identity.

14 Y. N. Harari, *Homo Deus: A Brief History of Tomorrow* (Harper Collins, New York, 2017).

15 For example, an Intergovernmental Panel on Climate Change report in 2018 suggested we only have 12 years to halve global greenhouse gas emissions to prevent global temperature rise of greater than 1.5°C, which are predicted to have significant negative impacts on human livelihoods and global wildlife.

16 T. H. Oliver, *Science* (2016) 353, 220–221.

17 M. J. Gelfand et al., *Science* (2011) 332, 1100–1104; J. Greenberg

et al. (eds), *Handbook of Experimental Existential Psychology* (Guilford Press, New York, 2004).

18 The International Migration Report estimates that in 2017 there were 258 million people who had migrated from their country of birth with 26 million of these refugees or asylum seekers.

19 T. Crompton and T. Kasser, *Meeting Environmental Challenges: The role of human identity.*

Chapter 19

1 N. Christakis and J. Fowler, *Connected.*

2 S. D. Sampson, 'Interbeing' in *This Will Make You Smarter*, pp 289–291.

3 This academic literature can be found grouped under the term of '*critical ecosystem transitions*'. Examples of ecosystems locked into undesirable states have been shown for forests, deserts, coral reefs, but the science of system transitions has application well beyond ecosystems.

4 Technofixes include ideas like geoengineering to prevent climate change and the development of robotic pollinators. Such ideas, although alluring, often carry high risks they will not work or will have unintended consequences, while others may be prohibitively expensive to move beyond prototype stages. For example, robotic pollinators have been developed, but producing the millions of machines needed to pollinate our crops would be hugely time and resource intensive. If only there was some kind of self-learning biological robot that could feed on renewable resources, and regenerate itself to create more robots . . . hold on, what about bees? The most effective solution to the pollination problem it to protect the amazing biological solutions that already exist. The same goes for the self-regulating climate system—the best thing is to not disrupt it irreversibly in the first place. Oliver et al., *Global Sustainability* (2018) 1. e9, ISSN 2059–4798

5 S. Ramgopal et al., *Epilepsy & Behavior* (2014) 37, 291–307.

6   Whereas the change in brain state in amateur meditators is more vulnerable – sudden stress events still cause perturbations – in advanced meditators relaxed brain states seem to become much more permanent and resilient. D. Goleman, R.L. Davidson, *The Science of Meditation*.

7   M. Gladwell, *The Tipping Point* (Abacus, 2002).

Further Reading

1   Wikipedia had 2,666,631 registered editors (who edited at least 10 times since they arrived) as of Dec 2018. https://stats.wikimedia.org/EN/TablesWikipediaZZ.htm#wikipedians.

Index